Dynamical Modeling of the Onset of War

Dynamical Modeling of the Onset of War

Alvin M Saperstein

Wayne State University, Detroit, Michigan, USA

World Scientific

Singapore • New Jersey • London • Hong Kong

Published by

World Scientific Publishing Co. Pte. Ltd.

P O Box 128, Farrer Road, Singapore 912805

USA office: Suite 1B, 1060 Main Street, River Edge, NJ 07661

UK office: 57 Shelton Street, Covent Garden, London WC2H 9HE

British Library Cataloguing-in-Publication Data
A catalogue record for this book is available from the British Library.

DYNAMICAL MODELING OF THE ONSET OF WAR

ISBN 981-02-4064-3

PREFACE

Nations have been making war upon each other throughout recorded history. Many people have been attracted by war's glory and gore; many others have been repelled by it. There have always been those who have attempted to "understand" war's origins and practice. Some wish to satisfy their intellectual curiosity about war and human conflict. Others have hoped to help frame policy so as to control war - to avoid it when possible, to win it or to minimize losses when desired or necessary. Consequently, the written record contains much descriptive and analytical discussion of the origins of war.

There have been two devastating "conventional" world wars in this century. The international system of nominally independent, competing nation states, which gave rise to these (and past) wars, just barely survived the vast "exchanges" of conventional high explosives. The possibility of a nuclear war increases the scale of explosive exchange by factors of millions; the intensity of their use (the rate at which explosives are detonated over civilians and combatants) rises by still further millions. The concomitant major global environmental changes raises the prospect that the outbreak of another major war may destroy the international system itself! Rather than simply modifying the strengths, numbers, and/or the identities of the actors which make up the system, as was the case in the past, a global nuclear war may eliminate the very system of nations which undergirds the possibility of war.

Thus curiosity about the forces leading to war, and the need to frame practical policy to survive as a nationby avoiding such a war, were intensified in the post-World War II period. Discussion of the topic of war initiation has become much more prevalent among the intellectual and political classes - those who have always been concerned with such issues. In addition, new groups, not previously heard on the subject, have registered their compelling interest in the question of the onset of war. There has been a significant display of concern with the issue from the general publics - the potential "victim classes" of a nuclear holocaust.

Much of this discussion on the onset of war has been traditional in form. There have been extensive verbal descriptions of "what" happened and verbal analysis of "how" and "why". No serious contention has been raised, nor will it be in this book, that this conventional approach is not valid or not useful. However, an alternative - mathematical - approach has been suggested by a number of scholars in the post-WW II era. Some examples of such a *mathematical political science of international security and war's onset* are developed in this book. The possibilities of applying qualitative and quantitative mathematical reasoning to the problem of international security and the origins of war are endless - perhaps as diverse as the variety of verbal analysis and description historically and commonly employed. The specific goal of this book is to develop the idea of "chaos" in a deterministic international system, and to apply it to simple mathematical models of the interactions between competing states in such a system. It is hoped to use the resultsof such model analysis, via an analogy between such "chaos" and the political notion of "crisis instability", in the framing of effective national security policies.

I emphasize that the methods and results obtained in this book are to be viewed as complementary to the usual approaches of verbal political analysis. If the results concur with those obtained in the usual fashion, it is presumed that our belief in these results is strengthened: understanding is enhanced, both theoretical approaches have increased validity, and policy based upon the results will be more likely to prove useful. If the two different approaches - verbal and mathematical - lead to discordant results, much more thought and development is required - *by each.* The fault may lie in the specific models used for analysis in each approach, the general typeology of model, or the basic approach itself. One great advantage of the mathematical approach advocated here, in contrast to the conventional verbal paradigms, is the explicitness of the assumptions made and the manipulations of these assumptions to produce conclusions. It is much easier to try new approaches and ascertain where they might have gone wrong.

Much of this book is based upon a research program which I initiated some fifteen years ago. It has been carried out primarily at Wayne State University, where I have been a professor of physics and an active member of its Center for Peace and Conflict Studies. It also included significant extended stays at the Stockholm International Peace Research Institute (SIPRI), the London International Institute of Strategic Studies (IISS), the Peace Research Institute of Oslo (PRIO), and the U.S. Arms Control and Disarmament Agency (ACDA). I am very grateful to my colleagues at each of these institutions. The program has resulted in a number of publications, many of which are mentioned in the reference sections of the following chapters. Chapter Five is largely based upon my paper "Chaos as a Tool for Exploring Questions of International Security" which appeared in *Conflict Management and Peace Science.* I am indebted to its editor, Stuart Bremer, for his encouragement of the original paper as well as his permission to use the paper in this book. For similar reasons, I am appreciative of Ray Dacey, who edited the special issue of *Synthese,* in which my paper "Mathematical Modeling of the Effects of 'Capability' and 'Intent' on the Stability of a Competitive International System" appeared; much of Chapter Six is taken from this paper, which is used with permission of Kluwer Academic Publishers. Much gratitude is due to the editors of the *American Scientist,* who helped me develop a "popular" version of my previous "scholarly" work ("War and Chaos", *American Scientist,* **83**, 548-557, 1995). It was this article which attracted the attention of Frank Ling of World Scientific Publishing Co. and ultimately led to this book. I am very grateful to the editors of World Scientific for their encouragement and publishing of this book.

I am also indebted to Prof. Fred Pearson, director of our Center for Peace and Conflict Studies for his continued encouragement of these endeavors of mine which, seemingly, have taken me out of the normal bounds of the research expected of a professor of physics. Finally, I have to express my continuing gratitude, and love, to the one who makes all of my scholarly activity - in fact all of my activity - possible: my friend and wife, Harriet Saperstein.

CONTENTS

CHAPTER 1

INTRODUCTION

At first glance, political science - and its subdivision, international relations - may not appear to have much in common with the science of physics. Actually, both sciences are concerned to some degree with the creation and dissolution of order and the interactions between components of a system. The physicist, for example, might ask how ordered crystals form from disordered fluids, how the crystals later fall back into the disorder of liquids and gases, or how nuclear particles interact as they scatter off of each other. The social scientist might be concerned with the formation of international alliances, their collapse into the disorder of war, or how nation states compete or cooperate in the international system. The strength of science, in contrast with other paradigms of knowing, is that it seeks many different, alternative routes to the understanding of the phenomenon of interest.

Scientific knowledge consists of multiple relationships between its many elements, multiple means of observing and/or measuring these elements. Hence, it may be useful to the development of political science to apply some of the methods of physical science - not in an attempt to supplant the conventional social science approaches but in the hope of providing alternative approaches to attaining its understandings and hence strengthening the belief in its results.

1. Desired Goals for this Book

In this work I hope to develop some aides to further the understanding of the world system of states. Specifically the aim is to further understand how states, and alliances of states, arise and disappear; from whence do they come? Into what do they go? How do they compete with each other - - in war and otherwise? How do they cooperate with each other - - via friendship, or coalescing into megastates? *We are seeking the laws governing the origin and dissolution of structure in the international system, and the wars which accompany and further these system changes.* Knowledge of such laws would enable the decision-makers of the nations concerned to make rational security policy, policies to ensure, and perhaps extend, the national security and wellbeing of the system's components.

By "rational policy making", I mean the following sequence of thoughts and actions: if I do a, then A will follow; if I do b, then B will follow; I prefer outcome A (or I find outcome B distasteful), so I will do a. An analysis is made of the different outcomes stemming from different policy options, and then a choice is made among these options. "I" in this discussion is not necessarily the unitary decision-maker at the head of a nation, deciding among different action "options" he/she will adopt towards other nations. "I" could mean the collective decision making apparatus of a nation - - limited oligarchy or unlimited electorate. The policy options include domestic actions required to obtain national consent to carry out the proposed international actions. Thus there is no need to restrict these ideas to "national elite" theories of international relations (B may be distasteful to the dictator or the man in the street; a may include domestic as well as international political options).

Note that such a rational policy making is of an "if... then...." format; it is dependent upon a necessary relation between one time and another, between now and the future -- a prediction. The situation confronting the policy-maker is that he/she may choose among a number of possible present options - - the initial variables. Each one of them has its own consequent outcomes - - the output variables - - related to the input by the structure of the scientific theory. The most desirable, or the least undesirable, outcome is attained by choosing the appropriate initial option.

Note also that the concept "desirable" belongs to the realm of ethical, political, cultural,...studies rather than to science per se. We are all aware that science has furthered ends usually considered evil (e.g., Nazi racial anthropology) as well as those universally regarded as "good" (e.g., the elimination of small pox). Of course, as a science and its associated technologies unfold, and the various possibilities inherent in a given situation become obvious, our notions of "desirable" may change (e.g., with the increasing availability of non-human energy, slavery may have become less "desirable" to many; so too, progress in the knowledge of human biology, may make racism less generally "acceptable"). Thus, great care must be taken in the enunciation of that which is "desirable". However, as we emphasize politics, etc., in the formulation of "desirable", we must be ever alert in our attempts to keep these forces out of the creation and criticism of the science itself. (Remember the story of Galileo's clerical opponents who refused to look through his telescope. They feared depriving the earth of its support at the center of the universe [scientifically bad!] because they wished to preserve a "man-centered" universe [ethically good!].) This entire "if... then.... " procedure is applied, of course, to a model of the system of interest. If the science is good and the model appropriate, the predictions for the model also apply to the real situation. The relation sketched, for the relation between policy and science, would be valid for the verbal sciences as well as the mathematical ones to be developed in the following chapters. The difference is in how the deduction of the outcomes from the initial premises - - the predictions - - are made: by qualitative verbal means or precise mathematical means. In all cases, the procedure is the same: for each possible option, use your scientific understanding of the model representing the situation confronting you to predict its outcome; then choose among your options. The alternative to this science-model-prediction approach is trial-and-error: pick an option, in ignorance of how it will turn out, and sit back to see what happens; if you don't like the way the situation is developing, you may be able to turn back before its too late. In the less dangerous world of the past, trial-and-error policy making was common; it's not clear that the nuclear world can afford many major errors in the future. It seems more reasonable to attempt to predict possible futures rather than to blithely await whatever comes.

The readers of this book are presumed to be graduate students in the early years of their study of political science, having some knowledge of the historical and political structure of the present "world system" as well as some insight into the structure and content of the various natural and social sciences. It is also hoped that their faculty, as well as present and future policy makers stemming from various academic backgrounds, will also find the book interesting and useful. No claim is made that this book represents the only scientific study of the "international system" or the first application of physical modeling to such systems. Indeed such studies have occupied the attention of philosophers, military officers, economists and business people, politicians, and political scientists ever since these categories existed. They have written much (see the end of this chapter for some suggested background readings); I have sampled a small part of this literature, and profited greatly thereby. My contribution, aside from the emphasis on the physicist's "way of thinking" (not unique to me; cf. e.g., Midlarsky in the "standard" political science literature), is the introduction of the concepts of "chaos" and "complexity" into the formal study of the subject. These concepts are seeing growing utilization (hopefully not faddishly) among practitioners and scholars in the field. Thus, in addition to an introduction to "scientific method", this book should be useful as a survey/introduction to the application of chaos, complexity, and the physicist's mode of thinking to the modeling of international relations.

2. Why Study International Relations Scientifically?

Human beings have a need to "understand" and to "control" their environment - physical, social, national, and international. Science is a means, proven successful by recent history, to attain these needs.

The need to understand seems to be universal and deeply rooted in human desires. It is often a prime motivation in human behavior and attitudes; people will go to great lengths to achieve "understanding". But this understanding can often have a superhuman or supernatural basis or entirety. People may sit at the feet of a"guru" for wisdom. Or they may directly invoke the role of Gods or other supernatural beings to explain behavior of interest. The planets may move through their complex courses in the heavens propelled by angelic movers. Thunder may be the bowling of elves. Trees may grow upward because of the desire of their inner spiritual nature to unite with heavenly spirits. Such "understandings" are obtained via direct revelation, or are revealed by some form of priesthood or society of mythmakers. They have been prevalent in the past and are still common today. I don't wish to slight such alternative modes of understanding but in this book we seek "scientific understanding" - the meaning of which will be teased out as we go along (although I do assume that all readers will have some intuitive notion of the phrase even at the start of this discourse).

The triumphs of the physical and biological sciences over the last few centuries as foundations for human control over our technological and medical environments are common knowledge. But the need is also often expressed for control over the political environment--national and international. Certainly sufficient control over the international system is wanted to avoid war-- when that course is desired, to win wars--when they are inescapable or seen as potentially beneficial, or to gain commercial advantages. Here too, this desire for contral over international relations has often sought fulfillment via the "unnatural". Magical incantations and/or prayers and supplications to some "higher beings" have been resorted to in order to force other individuals, groups, or nations to conform to our desires. Such approaches are still common, though they will not be further discussed in this book. Here, the interest is in <u>control based upon scientific understanding</u>, using our knowledge of the international system to manipulate that system to secure advantages or avoid disadvantages.

By "international system", I mean a collection of nations interacting with each other. The number and kinds of nations need not be fixed. Nations may be born, they may grow or decline, change significantly in character, they may die - disappearing from the international scene though leaving significant after-effects. (Dead nations are never just buried and forgotten!) Interaction between nations may have commercial, cultural, diplomatic, military, population exchange, and/or tourism aspects. Manufactured goods and technology, natural resources, energy, services, entertainment, information, and people, may be exchanged, as well as air, water, and land pollution. Interactions may be minimal--as appears to be the case, for example of Uruguay and Uzbekistan, or maximum - e.g., Canada and the U.S. appear to be very tightly coupled to each other. A major concern of this book is how these couplings modify the behavior, internal and external, of the nations concerned. It seems fairly obvious that the contemporary world system is much more strongly coupled than the worlds which we only know through history.

The scientific understanding, upon which we hope to base our control of our environment, differs from other "understandings" only in the process by which it is attained. Insights into that process--into the "scientific method"--are best obtained by viewing it in action, as will be attempted in this book. However, a brief sketch can be given in this introduction. The building of science is the weaving of a tightly knit fabric of understanding and control: hypothetical understandings are used to support possibilities of control; the success or failure of these controls--controlled experiments and observations--leads to refinements of the initial understandings or their discard for alternative ones; and the process is repeated, over and over again.

The use of the metaphors "fabric" and "weaving" is very deliberate here: no single "point" of factual or conceptual knowledge, in a science, is linked to other comparable points by a single thread. A science is only as strong as the multitude of ways in which one item of understanding can be reached from others. Hence, this book stems from the premise that application of some of the mathematical modeling methods of the physical sciences to the social sciences can only strengthen the latter.

Science can be construed as the search for testable understanding. Postulates--often, but not always, drawn from direct experience--are made about the behavior of some aspect of the environment to be understood. Deductions, either by formal logic, mathematics, or verbally, are then made, linking these hypotheses to some directly observable phenomenon in that environment. The required observations are then made--that is the "test". Failure of the observed values to agree with the deduced values implies shortcomings either in the observational process, the deductive process, or in the hypotheses themselves--some, or all, of which may have to be improved or changed to continue the process. Agreement between the observed and deduced values means that the process may be continued: one thread of the fabric has been tentatively laid down and the search for others may continue.

Two simple examples from elementary physics may be helpful. Starting with the intuitive notion of force, and a carefully defined acceleration concept, the hypothesis is made that the acceleration of a given body is proportional to the force applied to it. Given the acceleration, the distance traveled by the body in a given time can be deduced and then, under idealized conditions, simply measured. Changing the forces acting upon the body leads to changed accelerations and hence to changed distances. Thus the assumed proportionality between force and acceleration can be either confirmed or denied. Given this physical law, Newton's Second Law of Motion, a further hypothesis can then be made, linking the gravitational forces between objects to the distances between them-- Newton's Law of Universal Gravitation. Applying this law to the motion of planets, the future positions of specified planets in the heavens can be deduced, using the second law as an important part of the deductive chain, given their observed present positions. Testing these deductions by future observations of the planets is again used to confirm or deny the growing body of hypotheses.

As the science grows, not all deductions can be tested: some simply have to be taken as new knowledge or understanding in and of themselves, not directly confirmable or derivable in any way. For example, consider the science of spectroscopy as applied to astrophysics. Spectroscopy deals with the patterns of colors emitted by glowing hot substances: each chemical element displays its own characteristic pattern just as each specific musical instrument exhibits a specific pattern of sound. Knowing the element, and the conditions under which it is emitting light, one can deduce the expected pattern. Given an observed pattern, one can deduce the element and conditions responsible for its display. All of this is well confirmed by testing in terrestrial laboratories. Now apply it to celestial phenomena: the light from stars. From the light patterns observed in telescopes and other instruments, one can deduce the constituents and conditions of the stars and thus construct a science of stellar evolution. But--nobody has ever been to a star to sample its constituents and structure and it is extremely unlikely that such direct testing will ever be possible. Our understanding of astrophysics is only indirect and is strengthened as many indirect paths lead from initial understandings to further ones. Thus the sample model of a science--assume, deduce, test, assume, deduce, test,.... -- is far too limiting. When well tested in one realm, science is often used to extend our knowledge into non-testable regions. This will be the case for political as well as physical science.

So far, the deductions which have been mentioned have been predictions: given the state of the system of interest at the current instant and the hypotheses believed operant, the future state of the system, and the outcome of future observations in that system, are deduced: if...., then. But retrodiction, the deduction of past behavior from that of the present, is also important in the development of scientific understanding. If there is some means of confirming or denying the deduction, that too is a valuable test of the theory. For example, if we have a theory of human social behavior which can relate today's social behavior to that of the past, combining present day observation with the historical record leads to a fundamental test of that theory. Uncertainties in the history or the observations, for a given time interval and a given society, lead to uncertainties and hence weakness of the test. But the same procedure and theory would then be applied to many intervals, for many societies. Each single, "thread" may be weak, but the resultant fabric, of many such threads, may be quite strong.

Thus, science may be construed as a system of maximal relationships: each hypothesis supports many conclusions (you don't have one hypothesis with each fact to be explained; if you did, that would be an "*ad hoc*" hypothesis); each conclusion may be deduced via many different logical routes, starting from relatively few assumptions. This last statement can be rephrased as the requirement that there be a minimum number of *ad-hoc* hypotheses. This requirement - - a non-testable constraint on the thinking process, needed to make a science possible - - is often referred to as "Occam's Razor" (William of Occam, early 14th Century English scholastic philosopher): "What can be done with fewer assumptions is done in vain with more". Use as few hypotheses as possible to explain a given phenomenon or answer a specific question.

As an example of the use of this principle to cut away fruitless paths of inquiry, consider looking through the window of a house, from inside to outside, and asking the question: is there really an outside, out there, while you are inside? How do you know that what appears to be a window is not just a painting of an external world which may not really be there? Do a simple test: move parallel to the wall. If it is truly a window onto the exterior, its appearance should change. It does, but you may frame the hypothesis that there is a complicated mechanism, natural or supernatural, which automatically changes from one picture to another in accordance with your motion.

You can pile hypothesis upon hypothesis, so as to account for all possible observations you may make from inside, with the conclusion that you are merely looking at a "picture". It is much simpler to make the one hypothesis that you are looking through a window. You cannot test that assumption though you may then proceed to create an understanding of that external world based upon your observations. Occam's razor cannot be proved. It is an article-of-faith, a working hypothesis with which the human mind attempts to build understanding of its environment: start with the fewest number of simplest assumptions first; you may have to-- begrudgingly-- add complications or more later as experience -- testing – requires.

3. Science, Technology, and Policy Making

If science is the striving for testable understanding, technology is the striving for creation or control: to build homes, bridges, cathedrals, weapons, etc., to control pollution, markets, people, nations, etc. In order to build, we must know how to build, what works, what doesn't work, in the situation at hand. Such knowledge is often based upon custom: this type of cathedral structure worked in similar circumstances in the past; it should work here now. This customary knowledge is often the result of trial and error with minimal "understanding": this procedure failed in the past; we don't know why. Thus technology has always had a significant dependence upon history and will continue to do so. In the last few centuries, it has become more and more founded upon science: we understand the properties of given structures and materials and can thus say, a priori, whether or not they are suitable for the envisioned task.

Actually, the relation between technology and science is not the one-way dependence of the former upon the latter but a circular feed-back loop between the two. Understanding leads to the creation of new tools (new technology!) which can be used in further tests of the science. Technology leads to new questions and thus the need for further understanding. As common examples, consider lasers and computers. Lasers, built upon the pure science of the emission and absorption of light by atoms and molecules, are used for technologies such as surgery, the cutting of steel, and the playing of CD's. Laser technology is now used to extend the science of atoms and molecules by controlling the motions and states of these entities in predetermined manners. Computers, based upon microelectronic chips, stemming from previous understandings of the physics of solids, underpin much of the contemporary commercial, industrial, and entertainment worlds, and conveniently perform the computations which make further advances in physical understanding possible.

In practice, the forefront scientist may have to create new and superior technology to carry out the new tests required by the scientist, while the pace-setting technologist may have to attain new understanding in order to carry out the desired creations. The difference between leading scientists and trail-blazing engineers is often motivation - - to understand or to create - - rather than activity. And motivation is often not easily discernible.

In this book, we shall be much interested in policy making, which I take to be technology in the domestic or international political arenas. Policy making is to political science as engineering is to physical science and as medicine is to biological science. This policy making has also been primarily based upon historical understanding. However, unlike the situation in cathedral building or surgical repair, "history never repeats itself". It is much more difficult to find similarities between past and present domestic or international political situations comparable to those commonly found between past and present circumstances in the biological or physical environments. Cathedrals in Britain and in Serbia are much more similar in structure than, say, the responses of their respective peoples to the possibility of religious or cultural differences.

Small differences in the initial background of political events generally lead to larger differences in outcome than is the case in the non-political world. Hence it is much chancier to depend upon custom in determining political policy than it is in the other branches of technology. Furthermore, the political environment seems to be getting more dangerous: mistakes in formulating policy in a closely-coupled world filled with the knowledge of biological, chemical, and nuclear weapons will have far more serious consequences for humanity than was the case for "less accomplished" worlds. Clearly there is great need, as we extrapolate the lessons of the past to new political situations, for means to appropriately correct the lessons to account for the occasional major differences and ever-present minor differences in circumstances. In the absence of revelation from supernatural sources, such corrective means have to be sought in a scientific understanding of the world political situation. This book is a part of the general endeavor to create such an understanding.

4. Why Study Mathematical Dynamical Modeling?

All that has been said up to now would apply to conventional political science - a science expressed exclusively in words, and hence referred to here as a "verbal science". (It may very well include statistics in its data set, e.g., the number of wars in a given place or time, but the sought for relations between these numbers are expressed verbally. Science is the human-created set of relationships between the data, not the data themselves). The question here is why introduce dynamical modeling and why do it mathematically?

"Dynamics" refers to changing with time, with the understanding that there are causes for such changes. Our political world certainly changes over time and our search for "understanding" is a search for the "causes" of these changes. Knowing the causes, we may be able to alter them so as to produce changes more to our liking. Hence dynamical modeling is an important component of verbal political science. "Modeling" refers to the creation of a representation of the world of interest--in your mind, on paper, or in a laboratory. You cannot incorporate the entire real world in your mind whereby you can manipulate it so as to attain understanding of its dynamics (change some aspect of it and see how the rest changes). Hence you use a representation, a partial world, which you hope, contains the aspects of that world important for the behavior that you wish to understand. Whether or not that hope is justified will be determined by subsequent testing in the real world, of the results of your understanding of the model world. Again, modeling is a necessary characteristic of conventional political science, though usually informally and implicitly.

For example, suppose you wish to use Newton's Theory of Universal Gravitation to understand the evolution of a galactic system. The theory specifies the forces that exist between any two bits of matter, given their masses and the distance between them. Coupling this theory with Newton's Laws ("Theory") of Motion, we can determine how all of the bits of matter in the system will move, given their initial configuration. Since the bits move, the entire configuration will change in a calculable manner. We now create a model: an assumption as to how the matter of the galactic system of interest--assumed made up of bits of matter represented as geometric points, certainly not a complete picture of real matter--was initially configured ("initially" meaning at some epoch prior to the configuration of the galaxy as currently observed). Now apply the Newtonian Laws to the model and see how it evolves according to the theory. If the resultant configuration does not resemble the observed one, the model is no good. (Logically, it could be the theories which are inadequate, but they have presumeably been well tested in other situations.) If the calculated configuration coincides with the one observed, we have gained some understanding of galactic evolution--the observed galaxy *could* have evolved from the assumed model. Given agreement between model and observation, we don't stop; we continue with other models, seeking to find how unique this agreement is. The more unique the agreement between theory-model-observation, the more understanding we say we have of the situation.

Verbal political science is an old subject, going back at least to Plato. Its practitioners make verbal assumptions, do verbal deductions to reach verbal conclusions, and then test these conclusions verbally. Why attempt to do things differently? In addition to the desire for alternate routes between premise and conclusion - - for additional threads in the fabric of the science, there should also be a recognition of some of the difficulties facing a purely verbal science. There is a tremendous amount of redundancy and slack in verbal communication. There is rarely a unique relationship between a word and its meaning. It is very hard to be sure whether or not a verbal description accurately describes an observation, and to determine, if they differ, precisely where do the differences lie. For example, to say that the number of soldiers a nation has depends upon its population is a vague statement, hard to contradict To say that one quantity is proportional to the other, i.e., the doubling of one implies the doubling of the other, is a much more precise and testable statement. It is often easy to stretch the qualitative conclusions of a verbal science to fit the qualitative observations of the world being explained. In the process of testing a qualitative model, it may be difficult to find a contradiction to it in the qualitative description of the observed real world. Without the possibility of finding a contradiction between theory and observation, you have no test and hence no science.

Seeking for contradiction is the essence of the scientific process. Suppose, on the other hand, that a contradiction is found. The science is wrong and has to be modified. But where, in the qualitative process of hypothesis creation, deduction, testing, did it go wrong? It's very difficult to pin down precisely the source of the discrepancy between the verbal assumptions and the verbal testing of the real world.

The essence of mathematics is the preciseness of the relation between its elements. The meaning of the individual elements may be obscure (what is a point or a line?) but the relationships between them are uniquely specified. Moreover, the manipulation of these relationships in the process of deducing consequences from premises is transparent at all stages and maintains the uniqueness of the relations. The relation between the beginning and the end of the process is clear and precise if the mathematics is done correctly. Errors in process are comparatively easy to find and to correct. Even the difficulty of specifying the real meaning of a mathematical element (e.g., what is the reality behind the concept of GNP - gross national product?) is an advantage in a mathematical argument as compared to the conventional verbal development. In the latter, the relation between the model being used and reality is usually implicit and obscure; the very fact that, necessarily, a model is being used is often glossed over. The practitioner of mathematical modeling knows that a model is being used and explicitly sets out, at the onset, the connection between the model elements and those of the real world being investigated (e.g., "let x be the number of fighter aircraft possessed by a nation" - - an easily measured quantity, or "let y be the hostility of a nation towards its neighbors" - - hard to specify precisely but at least the concept is clear). All in all, the testing of a mathematical science should be much more straightforward and meaningful than the corresponding testing of a qualitative science.

5.Mathematical Methods of Getting from Now to Then: Dynamic and Stochastic rules

The options and outcomes (initial and final variables) of interest in this book will be represented numerically, e.g., the number of intercontinental ballistic missiles (ICBMs) possessed by a nation, its GNP, its imports, its hostility towards its neighbors (not so obvious to quantify), etc. Thus the scientific understandings developed in this course have the form of mathematical relations between input and output numbers. These relations are the "mathematical laws" being sought. In general, mathematical relations between numbers can be either deterministic or stochastic. Both types of law should be of interest to political science.

Deterministic laws are rules uniquely relating the values of its variables at one instant in time to its values at the next instant. For example, suppose you have a bacterium, growing in a Petri Dish affording unlimited food and waste elimination and sufficient space to allow indefinite growth without harmful crowding. There are thus no constraints on the multiplication of these bacteria which grow by bifurcating every minute. At the starting time there is one bacterium. One minute later there will be two. Two minutes later there will be four. The minute following that there will be eight, and so forth. Every minute the number of bacteria doubles - that is the rule. Give me the time, in minutes after starting, and I will predict for you - - precisely - - the number of bacteria present. ($N(n) = N_0 2^n$, where N_0 is the number initially present, n is the number of minutes since the start, and N(n) is the number predicted to be present after these n minutes.) This rule - - this prediction - - can be tested simply by counting the number of bacteria actually present in the dish.

Stochastic rules are probability relationships; they are rules which give probabilities. For example, suppose I toss a fair coin once every minute. After n minutes, I have tossed the coin n times. How many heads will I have seen after all of these tosses? I cannot say. I can only give probabilities: the probability of one head after one toss is 1/2, of two heads after two tosses is 1/4, of three heads after three tosses is 1/8, etc. (The probability of n heads after n coin tosses is thus $1/2^n$; the probabilities for lesser numbers of heads in n tosses can also be found, though not so simply.) Stochastic rules allow probabilities to be obtained with certainty, but probabilities are not so simple to test. The meaning of the probability being 1/2 for getting one head for one toss is that if you repeated the experiment -- tossing the coin once, observing if you get a head -- over and over again, the ratio of the number of heads observed to the total number of repetitions will get closer and closer to 1/2 as the number of experiments increases.

Thus the testing of the probability prediction requires an "ensemble" of experiments -- the repetition of the same observation on a large number of identical systems. If the law of bacterial growth previously mentioned were a stochastic law rather than a deterministic law, its testing would require the counting of the number of bacteria present on each of many identical dishes at identical times after identical starts. This may not be too difficult (it is done all the time in bacteriology labs). However, we know of only one world system so it would be rather difficult to find the ensemble of worlds (or of identical nations) required to test stochastic rules of international relations. Care will be required in applying stochastic understanding to international relations.

Note that policy making requires *prediction* whereas the testable mathematical relation is *deterministic*. The two concepts are not necessarily the same! *A predictive system must be deterministic but a deterministic system need not be predictive.* This important distinction may be clarified by reference to the three parts of Fig. 1, in each of which the horizontal axis represents the progression of time and the vertical axis represents the value of some variable x which stands for the numerical measure of some important descriptor of the system of interest. The desired prediction about this system is to find the value of x at some future time t, given its value x_0 at an initial time t_0. Assuming that each value of x is related to the immediately previous value by a unique rule - - that the scientific understanding being used can be expressed via deterministic mathematics - - the resultant time evolution of x will take the form of a simple curve, starting at the initial point (t_0, x_0), as illustrated in Fig. 1a. Each point in the trajectory gives the value of x at the corresponding time t; the curve is a complete, unique, precise prediction. (To use for policy making, consider two possible initial values, x_0 and x_0', and construct the two corresponding trajectories. See which of the predicted values you prefer at the future time t, and choose accordingly from the starting values).

To develop and use such a trajectory, e.g., for policy making, you must input a value for x_0, representing your knowledge of the present world. But such knowledge must be the result of a real measurement on the real world. Such real measurements, made by real human beings, are never perfect - - they never give a single value but a range of starting values, written as "dx_0", determined by the accuracy of the observations. (Careful measurements can reduce the size of dx_0, the uncertainty, but never to zero). Thus, instead of having a single initial value, as in Fig. 1a, a realistic situation may be represented by a range of initial values, as in Fig. 1b; instead of a single trajectory from the present into the future, the result is a large collection of neighboring trajectories - - a "flow". If the flow is closely bunched, as in Fig. 1b, the range of possible values of x at some future time t is roughly the same as the range at the start: the inaccuracy in our knowledge of the future is comparable to the uncertainty in our knowledge of the present. In such a case, knowing as much about the future as we do about the present - - given that our mathematical model is correct--we can say that we have made a successful prediction. (Of course, if we are in the process of testing a new theory for new understanding, rather than using a successful theory upon which to

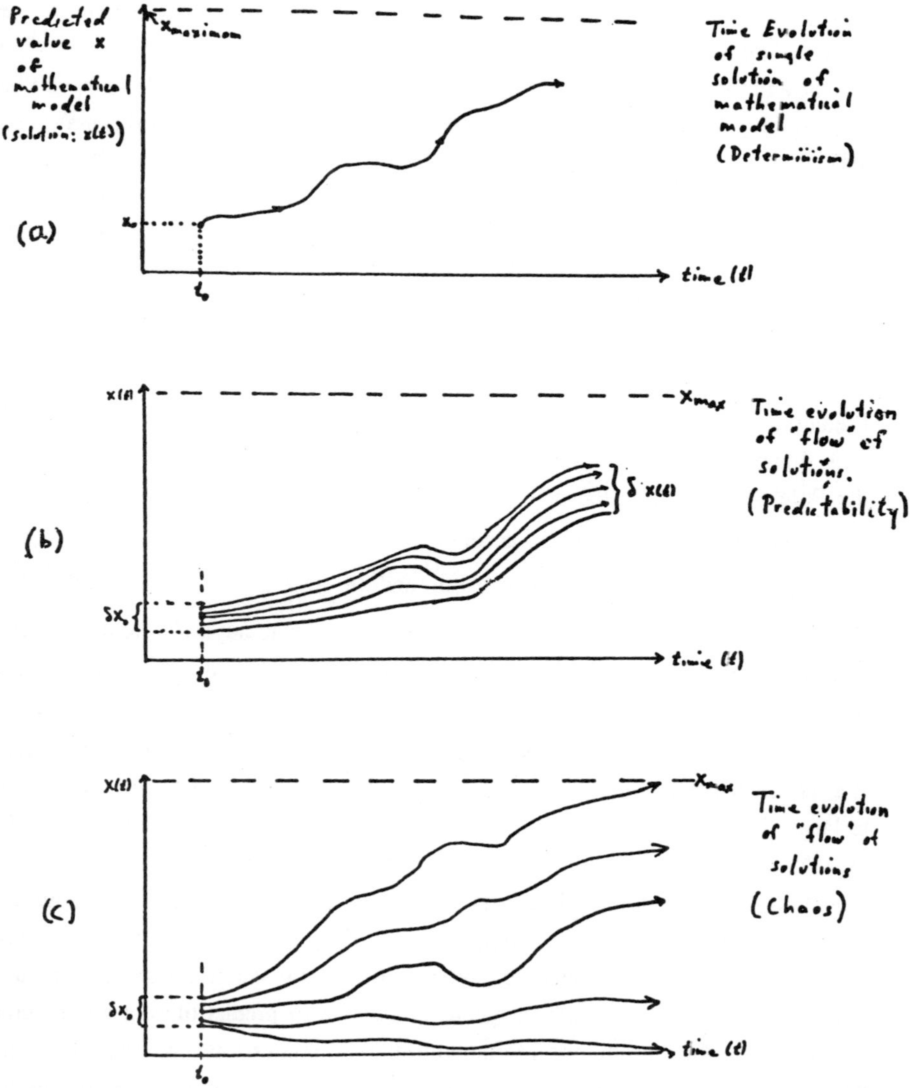

Fig. 1. Time evolution of the solution to a deterministic dynamical model. The solution starts at initial time t_0. (a) Single solution starting at the unique initial value x_0. (b) Set of possible solsutions starting from a range of neighboring possible starting values ∂x_0. The solutions remain neighboring with a range $\partial x(t)$ comperable to ∂x_0 so the model produces predictable results. (c) The solutions depart from each other, covering the entire range of values available to the model.. No predictions can be made with this model.

base policy making, we can then observe the actual future behavior of the real system, compare the results of these future measurements with the predicted values, and then accept, modify, or reject the theory, and the model we have used, accordingly). Again this predictive process may used to undergird rational policy making: choose the desired flow from among the different ones possible. (*The different trajectories are not allowed to cross each other*; if the deterministic mathematical system incorporates a unique rule, each value has only one possible successor value).

However, the flow may not be tightly bunched but may spread all over the allowed region, as illustrated in Fig. 1c. In this case, the range of future values includes all possible values, which means you know nothing about the future. Prediction is impossible so that neither testing nor policy making can be done with this theory. (For example, suppose you are trying to predict the angle from the Pole Star at which the planet Mars will be observed on some future night. You observe it tonight with an accuracy of 1°, that is, for example, you believe the angle to lie somewhere between 86^{0} and 88^{0} ($87^{0}\pm1^{0}$). If the width of the flow sometime in the future is 3°, you still know a great deal about the future position of the planet. If the predicted flow width is 360^{0} (the complete circle of the heavens!) you know nothing about its position in the sky: all you can say is that it is somewhere in the sky.) Such a situation is characterized as having "extreme sensitivity to initial conditions": a small change in starting values leads to large change in output values. A system manifesting such sensitivity is said to be a "chaotic system": no prediction of trajectory is possible.

At first glance, use of chaotic systems may seem useless from the point of view of understanding or controlling (making policy!) political systems. However, a given system may be chaotic under some circumstances, non-chaotic under others, and it may be very advantageous to be able to predict the conditions under which a transition from one regime (chaotic or non-chaotic) to the other may occur. For example, consider a fluid such as water. Under some circumstances, slowly and smoothly flowing in a wide, unobstructed channel, the water flow is "laminar": knowledge of the fluid behavior, at some instant of time at some point in the flow, allows easy extrapolation to knowledge of its behavior at neighboring locations at the same instant or at later times at the same location. The flow is predictable. Other circumstances, rapidly moving water in narrow, twisty, obstructed channels, lead to "turbulent" flow: knowledge of its behavior at a given time and place provides no insight into its behaviors at neighboring places or times. The flow is chaotic. The same water can flow from laminar regimes to turbulent regimes and back again so that any good scientific understanding of the behavior of the fluid should be able to represent both types of behavior and the transition between them. Practically speaking, the pilot of an aircraft wants the flow of the air fluid over his wing surfaces to be laminar; turbulent airflow may lead to loss of control. It is the job of the aircraft designer to tell the pilot what type of maneuvers to avoid so as to prevent the development of turbulence over the aircraft. In other words, the engineer must be able to predict transitions from predictable laminar flow to unpredictable turbulent flow in order to choose the appropriate policy for flying that particular aircraft smoothly, safely, and efficiently.

In a similar manner, later in this book, the dangerous situation in the international system, characterized as "crisis instability" by the political scientists, will be linked to the development of chaos in the model representing the system of nations. If crisis instability, with its concurrent high chance for the outbreak of war in the system, is to be avoided, then it would be very advantageous to know - - in advance - - which policy options led to chaos in the model representing the system, which did not. Thus the ability to predict chaos - to predict the transition to a regime of non-predictability - - should be very useful in formulating policy for the practice of international relations.

6. Testing the Scientific Theory

It's fairly easy to test a predictive - deterministic law. But success of the test does not mean that the law is necessarily correct nor does failure mean that it is wrong. To help understand this apparent contradiction, consider the following example. The law to be tested is the Newtonian Law of Gravitation; the model to be used is the usual solar system in which all planets, including our Earth, third one out from the Sun, are represented as points of definite known masses. The initial observation is the position of Mars tonight. The final observation - - to be predicted and checked - - is the position of Mars exactly one earth-year later. The actual deduction of the output numbers from the input numbers is carried out by computers programmed to carry out the appropriate operations ("integration") on the equations representing the Newton Laws of Motion and Gravitation. The process is carried out, the number obtained, a year passes, and the position is measured, after which the two numbers - - the predicted one and the observed one - - are compared. If they agree, i.e., the two bands of uncertainty for the two numbers overlap, the test is a success. But there may be other theories, which will predict the same number, starting with the same input. For example, Einstein's Theory of Gravitation will lead to the same results as obtained from Newton.

If they don't agree, if the two uncertainties do not overlap, something is wrong. But where is the error? Is the law itself wrong, or is it the model chosen in which to apply the law? Has there been an error in the parameters chosen to give specificity to the law, e.g., the value of the gravitational constant, the masses of the planets, or their initial distances from each other? Has the law been expressed appropriately for the computer and have the needed mathematical operations been carried out correctly? All of the above represent possibilities for error in the theoretical process. But the experimental process is also subject to mistakes. Have the initial celestial position measurements been done correctly and accurately? How good were the final observations? Thus a simple failure in the test does not immediately mean that a new theoretical understanding must be sought. Physical scientists have always been extremely conservative in interpreting the testing of their laws, new and old. It is suggested that the same conservatism should be characteristic of those seeking to understand international relations.

Turning to the testing of stochastic laws, ensembles are required. All atoms of a given type are identical, by definition, so there should be no difficulty in obtaining sufficiently large numbers of identical atoms to do the necessary statistical tests to check out probability predictions. Different bacteria are not identical by definition but there has seldom been any practical difficulty in acquiring large enough bacterial ensembles to carry out the appropriate testing. All cars of a given type, e.g., Chevies, are not identical -- they may have different colors. But for the testing of their driveability, these differences shouldn't matter so that, again, appropriate ensembles can be obtained for the testing of stochastic laws. But where is one to obtain ensembles of ethnic groups, nations, international systems? If the law states that the probability of conflict between ethnic groups differing in religion is twice as great as that for groups differing in language, how is this stochastic statement to be tested? This difficult problem will be bypassed for the time being.

How about testing predictions of the onset of chaos? If you can clearly characterize the observational difference between laminar flow and turbulence (which may not always be possible - there is "microturbulence"), test of the prediction should be straightforward. Once you are in the fully turbulent regime, behavior is seemingly random and thus ensemble - probability methods may be used - - for physical or biological systems; again, not for international systems!. Application to a political system may not be so apparent, transition to chaos in the model world may be a sign of crisis instability and/or the breakout of war in the real world, but can you be certain of the point in international relations when the real system becomes crisis unstable (except, perhaps, in hindsight)? Can you even identify the point at which war - - in contrast to armed clashes - - definitely begins? These questions will have to be examined again - - later.

7.How Many Variables Are Needed to Describe the International System?

From a practical point of view, you would like your theories - - of physics or politics - - to have relatively few variables. Even the biggest computers in the world cannot handle many accurately. Dealing with the motion of planets should not be so bad. Each planet moves in our common three-dimensional space; three position variables per planet times nine planets does not represent too much difficulty for a modern astronomical computer laboratory and leads to a fairly accurate description of our near astronomical environment. But here we are interested in an international system composed of roughly 100 nations.

How many variables to describe a nation of a million or so people? Even at one variable per person, that would be far too many to deal with productively; and how many people can be described by a single variable? Each individual can vary over many emotional and character dimensions, not to mention intellectual, biological, and economic dimensions. Even if such extremely detailed numerical information could be attained - - extremely unlikely - - there would be no practical way to manipulate the resultant millions of billions of variables. Must we give up attempts at a mathematical political science and stick to the conventional qualitative verbal understandings? (To be fair, a complete understanding should also encompass all of these variables!) Fortunately no - - there are clues as to how to productively proceed, again from physics.

Any ordinary sample of a simple gas contains billions of times more molecules than are numbered in the population of the earth. A complete description of any one of these molecules requires scores of variables. Hence the total number of variables required for a complete description of the gas is immensely greater than that apparently required to describe the international system. Yet, in practice, the specification of a very few variables is sufficient to allow all measurable properties of the gas to be successfully described and predicted. All that is required to completely specify the gas for most purposes is knowledge of its volume, pressure, temperature, and perhaps a few more partial pressures. Each of these few "macrovariables" (specifying the state of the gas as a whole - - on a human scale) is a suitable average over the values of the countless "microvariables" (specifying the behavior of the individual molecules). Common observations of the gas, and its dynamical evolution, give results which are really averages over the randomly fluctuating microbehaviors of its individual molecules. Ordinarily, the fluctuations of the overall behavior of the gas away from its average values are averaged out and hence not seen. If the gas sample is extraordinarily small, so that it includes very few molecules, the fluctuations become important and the description of the gas in terms of a very few variables is no longer successful.

Jumping to the social sciences for a moment, for a different but similar analogy, consider life insurance. The insurance company does not know when any given individual will die; but it does know the average number of deaths expected at any one time and sets its rates accordingly. Given a large population, deviations from the average are unimportant and the company will do as expected. In a small population sample, there may be large fluctuations from the expected averages in which case the insurance company may either make better than expected profits or go bankrupt.

Getting back to the gas analogy, the individual molecules collide with each other occasionally but the great preponderance of each molecule's time is spent very far away from, and hence independent of, its fellow molecules. Hence each molecule's behavior is determined by its own characteristics, independent of the characteristics and behavior of the other molecules of the gas. It's irrelevant to the individual molecule where it is in the gas or what values the macrovariables describing the gas have. We can talk about the properties of the individual independently of those of the collective. Since the behavior of each molecule, except for its very rare collisions with other molecules, is independent of whether or not it is contained in the gas, it follows that the properties of the gas - - the collective - - is just the sum of the properties of its individual member molecules. Hence, knowing the properties of the individual units, the properties of the collective may be predicted.

This approach to understanding is often called "*reductionism*": knowledge of the microenvironment directly gives a corresponding knowledge of the macroenvironment. Thus a full understanding of the behavior of the atomic nucleus leads to an understanding of the atom. Understanding the atom leads to an understanding of the molecule, which in turn leads to understanding of larger units and so on up to an understanding of the entire physical (and biological and social?) universe. An understanding of our grand and glorious universe is *reduced* to an understanding of its most elementary parts. A full believer in reductionism would thus say that it is only worthwhile struggling to understand the nucleus and its constituents. All the rest is derivative/secondary knowledge.

This of course assumes that it is possible to make the appropriate deductions that will take you up the chain, e.g., that a chaotic sensitivity to initial parameters will not prevent you from attaining reasonable knowledge about the higher levels of being. It also assumes that the whole is merely the sum of its parts - that the essence of each part is independent of the existence and behavior of the whole. Such an assumption seems to be valid for a diffuse gas. It is not true for all of the circumstances in which a molecule may find itself. Take the same molecules and pack them tightly together, e.g., condense gaseous water vapor into solid crystalline ice, and the behavior of each molecule is profoundly altered by the close proximity of its neighbors - - behavior of the one is sensitively dependent upon the nature and behavior of its surroundings. Microbehavior is not independent of the macroenvironment and hence the whole is not simply the sum of its parts.

Turning from a physical system of molecules to a political system of nations, analogy leads to the possibility of describing the latter by a relatively few variables, each stemming from appropriate population averages over the constituent nation populations. In subsequent chapters I will try to describe the international system in terms of variables represented national "intents" and national "capabilities". Intent variables will be averages over individual intents - - the psychosocial variables characteristic of individual members of the population - - what they want to do. Capability variables would represent averages over the economic and military resources available to the individuals - - what they are capable of doing.

For example, a variable representing the total income of the nation would be the sum of the incomes of each individual member of the nation. But, the income of an individual is not independent of the national income; the total economic climate has a major influence on the income of any citizen. And, e.g., the attitude of any one individual towards the outside world is certainly influenced by the corresponding national attitudes. Individual attitudes and behaviors are greatly influenced by national and international behavior and attitudes. Extensive observations of the behavior of an isolated human being leaves one ill-prepared to predict the behavior of that person in a social setting: family, club, nation, etc. People are more like the molecules of a solid than like those of a gas! [a] We would like to say that properties of the collective are suitable averages over parameters characteristic of the individual - that nations having specified capabilities and intents (uniquely determined by their populations) will behave in particular manners *vis-a-vi* nations with other values of these capabilities and intents. But how can we do this if the values of the parameters describing the individual depends upon the state and behavior of the collective - on averages over those very individual parameters themselves? As the collective changes, so too does the individual. So, how can we predict the behavior of the former on the basis of the behavior of the latter?

At first glance, it appears that the paradigm of "reductionism" - the idea that the whole is just the sum of its parts (presuming, as it does, that the parts exist independently of the whole) - is untenable when applied to a study of international relations. The opposite of reductionism might be characterized as "*holistics*"-- that you cannot understand the whole in terms of its parts but must create an entirely different science starting from the whole (e.g., developing a science of the physical behavior of gases without presuming the existence of molecules - a science which actually has been done [thermodynamics], with limited success). The practitioners of reductionism would counter that if you really know the properties of the individual, including knowledge of how that individual is governed by the environment in which it finds itself, that you could deduce the properties of the collective. Such an attitude towards a science, such a working hypothesis for the acquisition of further knowledge, is non-testable -, as is Occan's Razor. You can't say reductionism is wrong; its supporters will say you haven't pushed it far enough. The same can be said for holistics. However, when progress towards knowledge in a reductionism sense is made (as was the case with the physics of gases), it is usually felt that more understanding has been gained: you know more about something when you can link its behavior to that of its elements. I choose to try to make progress in this sense.

---

[a] This dichotomy is mirrored in traditional political science by the "rational choice" vs "adaptive behavior" dichotomy (Axelrod,1997, p.4). In a "rational choice" model, the individual follows system rules which are valid whether or not other individuals are present. In "adaptive choice" models, the individual must adapt its behavior under the system rules to the presence and behavior of other individuals in the same system.

But how is progress towards understanding the behavior of nations on the basis of that of their citizens to be made, given the conundrum that the behavior of individuals is based on the behavior of their nations just as national behavior stems from that of its member individuals. A possible means out of this impass is the notion of "*time scale*". We know that individual and national behavior may change over time but we shall assume that the scale of these time changes is very different: that fundamental human behavior changes, if at all, over time spans of generations whereas national behavior may change over the course of years. After all, Frenchmen are still French and Germans still German, whereas the enmity between their two nations has changed to comity in the course of a very few years during this century. There is thus some evidence to support the *heuristic* assumption that national changes - of the sort of interest to international relations - take place over intervals much shorter than those characteristic of human changes - in which case, the people parameters can be taken as constants during shifts of national behavior. Thus individual behaviors can -- hopefully to a good approximation -- be taken as existing independently of their evolving national environment and so the national environment can be expressed as simple averages of independently determined individual characteristics.

The result would be that nations, in their interactions with each other, would be describable in terms of relatively few parameters and variables, just as in the case for gases, for many (though not all!) purposes. Then the system of nations would evolve according to simple dynamical rules governing the behavior of these variables and based upon these fixed, empirically ascertainable, parameters. Following this dynamical evolution would allow the outcomes of various policy options to be predicted.

This is a heuristic approach with no *a priori* guarantee of success. All we can do is to try it and see if it works. This of course again raises the question of the meaning of "works". If the approach leads to an increase in understanding -- if its results are coherent with, and flesh out the body of, the understanding attained from the more traditional approaches to international relations; if the approach helps to shape productive policy in the international system -- by allowing policymakers to anticipate the potential consequences of their policy choice and choose accordingly; then the approach, as developed further in this book, will be said to "work".

8. Complexity — Unexpected Emergence of New Structures

We have seen that deterministic models may exhibit "extreme sensitivity to initial conditions," making prediction of specific behavior impossible. This general concept has been labeled as "chaotic" and illustrated by the behavior of turbulent fluids, wherein the behavior of the fluid at neighboring points in space and/or time is seemingly independent of the behavior of the "initial" point, no matter where and when that initial point is chosen in the turbulent fluid. It is also possible that such a theoretical model may exhibit unanticipated structures via a phenomenon called "bifurcation": the system evolves smoothly and predictably along a well defined path (such as found in Fig. 1c) until it reaches a point of extreme sensitivity; the system then transitions to one of several possible paths, representing different behaviors or structures — you don't know, ahead of time, which one (predicting fails!); it then proceeds on the newly chosen path, again smoothly and predictably. The overall apparent picture is illustrated in Fig. 2a, where again x is a numerical measure for some important descriptor of the system and t represents the time.

Since the system is presumed to be governed by deterministic laws, and such laws specify sequential behavior uniquely, the actual behavior must be as illustrated in Fig. 2b. Each path is uniquely sequential, continuous, and single valued. The choice of final path is contingent upon which of the two initial paths the system was actually following. The two paths before the bifurcation point are so close together that their difference is hidden by the finite errors made in the initial observations of the system — they are indistinguishable to any human observer. Except for, perhaps, probabilities, you don't <u>know</u> which of the two paths the system is "actually on." After the bifurcation point, the two paths diverge rapidly so that it soon becomes apparent which path represents the future evolution of the system — prediction is again feasible. Thus "bifurcation" (or "trifurcation," etc.) is another manifestation of the "extreme sensitivity to initial conditions" which may be present, under some conditions, in deterministic systems. The generalized sensitivity, including both bifurcation and chaos, is referred to as "complexity."

A complex system may exhibit predictable behavior and structures; it may also manifest unpredictable behaviors and structures, sequences of events and structures which depend upon previous events and structures in an apparently random manner — effective descriptions and explanations require the application of

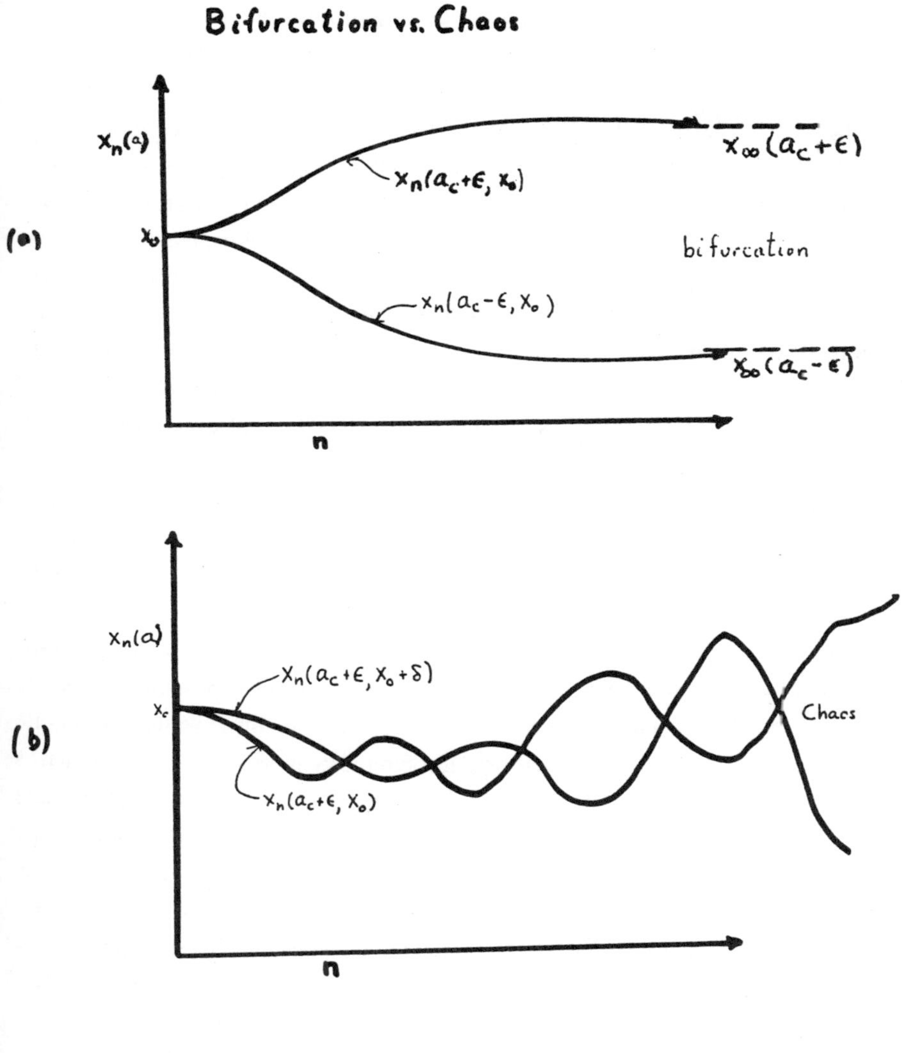

Fig. 2. Bifurcation versus Chaos. (a) The system is extremely sensitive to initial conditions at n=0. Slight changes about the starting value x_0 lead to grossly different out comes via alternate orbits starting at x_0. But the orbits remain well separated after departing from x_0, making it possible to determine which path the system is following. (b) the different orbits starting in the vicinity of x_0 repeatedly cross each other, randomly, after leaving x_0. Hence it is never possible to determine which trajectory governs the time evolution of the system. Prediction of outcome is impossible.

probability concepts. The ideas behind the display of both predictable and chaotic behavior by deterministic physical and social systems have been introduced. It is now appropriate to examine the possibilities of structure formation. In physics, this might be the growth of well-ordered crystals from previously free molecules. Socially, this might be the growth of towns from a previously freely dispersed rural population, the growth of cities and nations from among these towns, the growth of collections of nations in the forms of alliances, empires, etc.

The predictable growth of social structures is fairly straightforward and understood: a king or government decrees that a city shall be created in a given location; its city plan is laid out beforehand, and then it is built. The origin and initial construction of Washington, D.C. as the future capitol of the United States is a good example of such a process, leading to anticipated social structures.

To gain some insight into how unanticipated structures can arise in physical, biological, psychological, or social systems, even though such systems may be governed by deterministic rules, consider the following analogy. Assume a circular table, set for dinner. There are dinner plates symmetrically placed around the circumference of the table, and midway between each pair of plates is a wine glass. Nothing is specified as to whether the glass to the left or right of a given plate belongs to that plate. The initial symmetry is that either is possible. Once a person sits before one plate at the table and selects one of the two optional neighboring wine glasses, say the left one, the symmetry is broken. The only way the table can be filled is if all subsequent people sitting at the table also select the left-hand glass; the table becomes "left-handed." A small initial contingency — a choice — puts the table into one or the other of the two very distinct classes.

Assume now that many people are milling around, waiting to be seated for dinner. Some of these people are indifferent to the handedness of the table they will sit at. Others will only sit at a left-handed table. The remainder will only sit at a right-handed table. The three classes of people making up the banquet are randomly distributed throughout the hall. The *a priori* symmetry is that people of any of these classes are equally likely to be found anywhere. After the invitation to be seated, the symmetry is broken, and a structure is created in which people of the same class are clustered together. Suppose further that tables are waited on only if two neighboring tables are of the same handedness; the rest go hungry. The original crowd of people is now structured into two sets: one hungry, the other satiated. But one cannot know in advance who, if anyone, will go hungry; one has

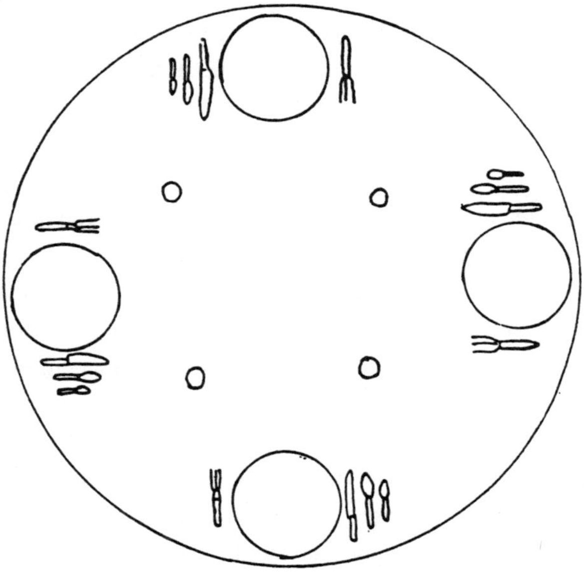

First person to reach for wine breaks the symmetry.

Fig. 3. Looking down upon a dining table, formally set for dinner. To which setting, left or right, do the wine glasses (small circles) belong?

to go through the process, step by step, to find out. Moreover, if the process were to be repeated under identical circumstances, different sets of people would be selected to go hungry.

The analogy could be extended and made more complex. For example, suppose that collections of three or more neighboring tables of the same handedness are served from large pots of food that are kept sanitary. The remaining people who are served, are served from small pots that harbor dangerous bacteria. The result is a new structure of ill and healthy people, arising from the original simple contingency. (There could, of course, be a series of contingencies; for example, only some of the small pots, randomly distributed, are contaminated.)

Each step in the model is deterministic. It is rigidly connected to the previous step by rigorous rules. The randomness of the input — who initially sits at each table — determines the outcome, which cannot be foreseen. The process of getting from the initial unstructured state to the final structured one, in the deterministic system, is dependent on the path, resulting from a small random element within the model or in the interaction between the model and its environment. It is important that the system be very sensitive to the small random element, that it "bifurcate" into one of the two very different paths as a result of the small disturbance. This sensitivity, which makes contingency possible, can come about only if the mathematical model describing the system of interest is inherently nonlinear — that is, if the changes in the dependent variables are not simply proportional to changes in the independent variables.

A physical example of the application of this "sensitivity" or "symmetry breaking" analogy occurs in the description of change of state, for example, the freezing of a gas into a solid as would occur when water vapor in the atmosphere turns into snow. The elements of ordinary matter are molecules. In the usual gas, these molecules are randomly distributed throughout the three-dimensional volume they occupy. Any point in the space is equally likely to be momentarily occupied by a molecule. There is no correlation between the molecules; knowing that one point in space is occupied tells you nothing about whether a different point is also occupied by a molecule. There is no "structure" to such a gas. The space is perfectly symmetric — every point is as likely to be occupied as every other one.

Consider now a crystal lattice of molecules resulting from the condensation of this gas into a solid. All points of space are not now equally likely to be occupied by molecules. If you know where some molecules are, you know the distances and directions to the location of others. Points not at these locations will not be occupied. The symmetry of the space is broken. You may not know where the first few molecules will locate themselves as the gas begins to condense (the initial space is still symmetric), but the location of all subsequent molecules condensing into the crystal is preordained once the initial ones are fixed.

The same paradigm is easily demonstrated in current ideas of biological growth. The contingency paradigm assumes that each cell contains a complete set of rules as to how it should respond to each potential environment (internal as well as external). It will behave one way if it finds itself surrounded by liver cells, a different way if surrounded by bone cells, and so on. Which specific environment it finds itself in depends on the "path" it has taken since the "beginning." At each stage along the path there will be random small disturbances as well as deterministic major rules. The sensitivity at any point, which will vary from point to point, will determine whether and which bifurcations occur. In this way biological organisms grow with the similarities and differences commonly observed.

Or for a political science application of the same concepts, consider groups of people spread uniformly over an extensive landscape. No location is favored; symmetry exists. But then towns spring up. People move to them. You are more likely to find people in the towns than at other locations, so the symmetry is broken, and instead, a structure has been impressed on the space (e.g.,Manrubia and Zanette,1998). Some of those towns eventually amalgamate into cities, and a different structure is now evident. Later still, the collections of cities have become nations that in turn form the structure we call the "international system."

A specific contingency model, illustrating the growth of alliances and/or supernations from an initial collection of nations is due to Axelrod (1997, Chapter 2). Starting with resources randomly distributed among a set of independent 'actors", a sequence of random demands is made, by one upon another, for tribute, following simple explicit criteria. Each demand is followed deterministically by acquiesense and growing subordination or war and mutual destruction of resources. Running many trial sequences on a desk computer, some runs lead to a smaller group of "supernations" with growing prosperity. Sometimes these new nations collapse into impoverished subgroups as the sequence continues. and sometimes they never form, the elements continuously dissipating their resources into internecine strife. Other examples of social structures evolving from sequences of contingency rules are to be found in Epstein and Axtell (1996).

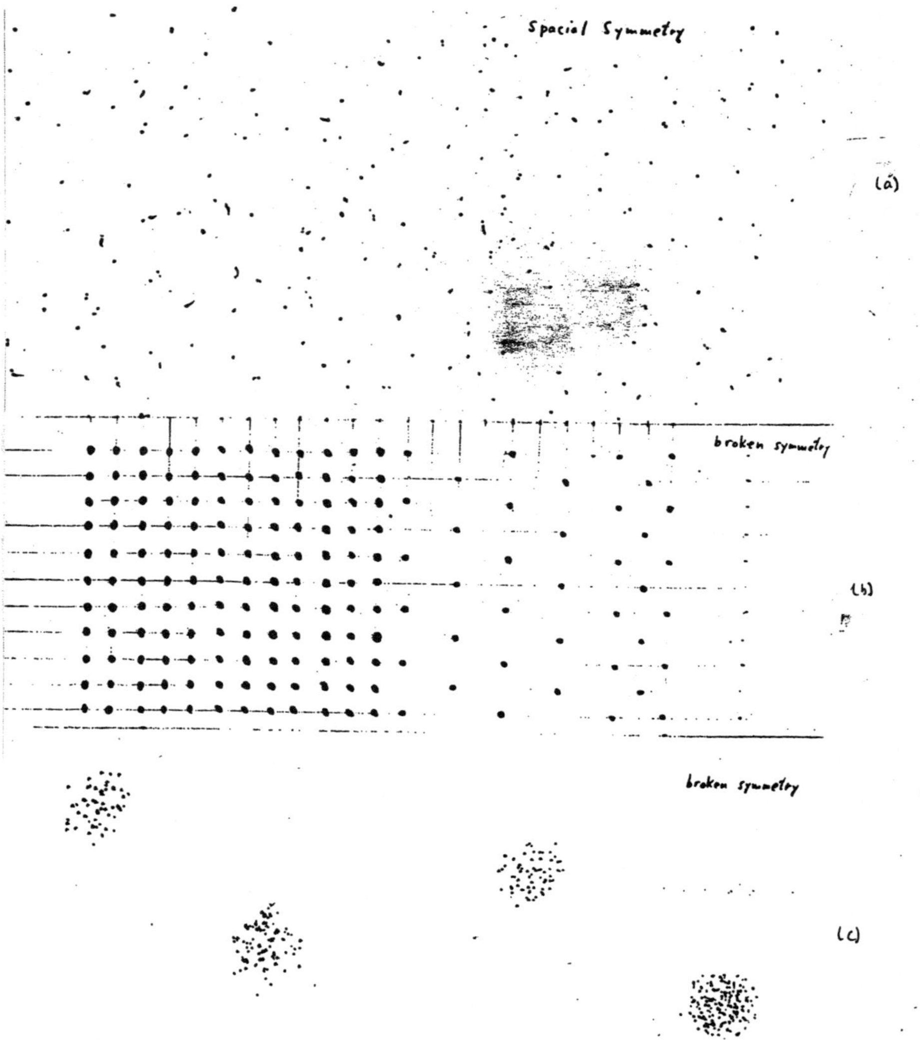

Fig. 4. Spacial Symmetry. (a) Each point in space is equally likely (or unlikely) to be occupied by an object (black dot). (b), (c) Broken Symmetry. Some points in space are more likely to be occupied than others. The result is either a patterning or clustering of these objects.

9. History

Examination of the above analogies quickly establishes the importance of history in complex systems. In order to understand how a system (physical, social, etc.) evolved from some initial configuration to its presently observed one, we must know its configuration at each intermediate instant. In computing the system's evolution, via some theoretical model, its configuration at every intermediate time must also be calculated, even though only the final state may be of interest.

This approach is very different from that of the simple, usually linear, physics taught at the elementary level. "Linear" means that changes in the system, produced by disturbances external to the system, are proportional to the size of the disturbance: small changes in input result in small output changes. Hence there can't be the "extreme sensitivity to external disturbances" characteristic of chaos and complexity. Linear systems are not complex. The physical laws applicable to such simple systems are usually integrable: the application of well established mathematical procedures to the initial configuration of the system produces the configuration at any subsequent time without the necessity of computing the intermediate steps. Small changes in the initial state of the system lead to correspondingly small changes in the final state. Hence the deterministic system is predictable. (Conversely, complexity may be defined as the set of deterministic theories that do not necessarily lead to long-term predictability.) Though small "perturbations" (small disturbances to the system which are not, usually cannot be, included in the theoretical laws used to calculate the evolution of the system) are always present at every intermediate stage in any real system, their impact is ignored as the calculation goes directly from beginning to end state. The actual history of the intermediate evolution of the system is not of great interest.

Complex systems are sensitive to small disturbances and hence cannot ignore them at any stage. Hence their evolution from beginning to end cannot be computed without actually calculating each intermediate configuration. Complexity theories thus depend on the complete "path" taken by the system between its beginning and end points — its history. Every intermediate instant of time may see the theoretical system diverted from the path it might have taken in the absence of the, always present, perturbations. Hence minute changes in the input parameters may lead to large, incalculable changes in the output. Prediction is no longer possible. The system is extremely context-dependent. The complexity paradigm should thus be very suitable for exploring the history-dependent social sciences.

It is clear that the international system is a history based system: everything that happens seems to depend upon all that has gone before. It is a contingent system: precise, long-range predictions seem to be impossible; the best that can be done is to say which outcomes are more probable than others. What isn't clear is whether this observed contingency is based upon underlying deterministic or stochastic laws. If the former, then there are rules connecting successive configurations of the system of nation states: some short-term prediction is possible, some order is expected. The question can only be addressed by creating theoretical models of the system, exploring the model behaviors, and comparing with the behaviors observed in the real system. That will be the task of the rest of this book.

The similarity between the calculated behavior of complexity-dominated systems and the behavior of sociopolitical systems in the real world gives considerable credence to the idea that the real world is dominated by deterministic rules and that the observed contingency is due to the occasional sensitivity of the real system to minor, but always present, random perturbations. This paradigm is quite different from that which supports the observed contingencies of the world on an underlying stochastic foundation. The choice between the two approaches to sociopolitical reality is not purely academic but has profound practical consequences. Both paradigms rule out the possibility of long-term prediction, but the complexity scheme does allow for short-term prediction and thus offers the possibility of control. If the ruling outlook in a population is that sociopolitical life is based on the stochastic paradigm, there is no point in political activity — in trying to form and direct collective behavior towards predetermined ends. The complexity paradigm makes it reasonable for a population to expect effective government and to believe in the possibility of a collective solution to a collective problem. Public-opinion polls in contemporary America seem to indicate that we are not convinced of the applicability of the complexity model to public life, these days, but the evidence is not yet all in.

10. References and Additional Readings

Axelrod, Robert 1997. *The Complexity of Cooperation.* Princeton, NJ: Princeton University Press.

Cambel, A.B. 1993. *Applied Chaos Theory: A Paradigm for Complexity.* New York, Academic Press.

Cioffi-Revilla, Claudio A. *The Scientific Measurement of International Conflict*

Dougherty, J.E. and Pfaltzgraff, R.L. Jr. 1990. *Contending Theories of International Relations: A Comprehensive Survey,* 3rd Edition. New York: Harper & Row.

Epstein, Joshua M. and Robert Axtell,1996. *Growing Artificial Societies.* Washington, D.C.: Brookings institution Press.

Ferguson Y.H. and Mansbach R.W. 1988. *The Elusive Quest: Theory and International Politics.* Columbia, SC. University of South Carolina Press.

George, Alexander L., 1993, *Bridging the Gap: Theory and Practice in Foreign Policy,* Washington: United States Institute of Peace Press.

Isard, Walter *Understanding Conflict and The Science of Peace*

Johnson, N. 1989.*The Limits of Political Science.* New York: Oxford University Press.

Manrubia, Susanna C. and Damian H. Zanette 1998. "Intermittency Model for urban Development", *Physical Review* **E 58**, 295.

Meehan, Eugene .J., 1971 *The Foundations of Political Analysis; Empirical and Normative.* Homewood, IL: the Dorsey Press.

----, 1968. *Explanation in Social Science: A System Paradigm.* Homewood, IL: The Dorsey Press.

----, 1994. *Social Inquiry:Needs, Possibilities, Limits.*Chatham, NJ: Chathem House Publishers.

Most, B.A. and Starr H., 1993. *Inquiry, Logic, and International Politic.* Columbia, SC: University of South Carolina Press.

Muncaster, R. & D. Zinnes "Structure and Hostility in Internati.Systems", *J.Theo. Politics* **2:1**, 31.

Nagel, E. 1961. *The Structure of Science.* New York: Harcourt, Brace, and World.

Nicholson, Michael 1992. *Rationality and the Analsis of International Conflict.* Cambridge: Cambridge University Press.

Rochester, J.M., 1992. "The Production and Transmission of Knowledge About International Relations: The Disheveled Discipline", in *Teaching World Politics: Contending Pedogogies for a New World Order.* eds. Lev S. Gonick and Edward Weisband. Boulder:Westview Press

Saperstein, Alvin M.1995."War and Chaos", *American Scientist,* **83** 548, Nov./Dec.

----,1997. "Complexity, Chaos, and National Security Policy: Metaphors or Tools?" in *Complexity, Global Politics, and National Security,* eds. D.S.Alberts and T.J. Czerwinski. Washington, DC: National Defense University Press.

Singer, J. David 1968. *Quantitative International Politics; Insights and Evidence.* New York: Free Press.

Singer, J.David, 1990. *Models, Methods, and Progress in World Politics.* Boulder: Westview Press.

Singleton, R. Jr. Straits, B.C. Straits, McAllister, R.J. 1988 *Approaches to Social Research.* New York: Oxford University Press.M.M.

Waltz, Kenneth N. 1979. *Theory of International Politics.* Reading, MA: Addison-Wesley.

Young, Jefferey R. 1998. "Using Computer Models to Study the Complexities of Human society", *Chronicle of Higher Education,* July 24, A17-20.

Zinnes, Dina A. 1976. *Contemporary Research in International Relations; A Perspective and Critical Appraisal.* New York: Free Press.

CHAPTER 2

PREDICTIVE STRUCTURE BUILDING — STATIC AND DYNAMIC

1. Static vs Dynamic Modeling in International Aggregation Studies

In recent years the professional political science literature has contained many articles on mathematical model building. The international relations and peace science communities have devoted much effort to creating, validating, analyzing, and using, mathematical models of nation and alliance formation and dissolution, international competition and arms races, battle and the processes of war, etc. The models, using the methods of discrete and/or continuum mathematics, have been resorted to by those interested in making and carrying out policy, by scholars, and by teachers. Though the emphasis in this book is on dynamical models, the reader should be aware that static models are also common. (Some examples of each are listed in *Table 1.*)

Static models (e.g., Kaye and Lindsey,1970; Intriligator and Brito,1984; Axelrod and Bennett, 1991) compare alternative configurations of the system being modeled using fixed criteria of superiority. The goal is a configuration having maximum gain or stability, or minimum "distance" between constituent entities, these terms ("gain"...) being suitably quantitatively defined. But nothing "happens" in such a model; if the system is initially in some configuration, it remains there; the transition process between different possible configurations is not part of the model. If one assumes that transition processes in real world systems require the passage of time, then static models are characterized by the absence of time as a system parameter or variable. In dynamic models (e.g., Richardson 1960a,b; Lee, Muncaster, and Zinnes 1994; Saperstein 1991, 1992a,b), the passage of time, or some other representation of the evolution of the system from one configuration to another, is a paramount independent variable. The output of such a model is an "orbit" in the "configuration space" of the system, a sequence of allowed configurations, extending into the indefinite future from an arbitrary "now" (e.g., Saperstein 1990).

When used as a predictive tool, static models presume that the system, will end up in the "most desired" configuration; they don't say how it will get "there" from "here". A change in the model represents a change in the relative desirability

Some "Typical" Mathematical/Physical Modelling Approaches to International Security

a. model variables
b. mathematical method
c. definition of stability
d. types of questions answered

Static Models	Dynamic Models
1. Kaye and Lindsey: (1970) a. No. of warheads for each of 2 parties b. Graphical display of inequalities. c. Stability area on 2-D plot where neither party's attack inequality holds. d. Crisis stability *	1. Richardson: (<1960) a. National weapons stocks b. Coupled linear differential equations c. Eigenvalues of differential-equations-matrix indicates exponential rise or fall of arms levels. d. Arms race escalation.*
2. Brito and Intrilligator: (1976 - 84) a. No. of warheads, number of casualties b. Coupled differential equations c. Inequality regions on 2-D plot d. Crisis stability *	2. Lee-Zinnes-Muncaster: (1991) a. $r_{ij}(t)$ "relationship" between i and j b. Coupled non-linear differential equations representing "structural interaction rules." c. Constancy of signs of r_{ij} (t) d. Time evolution of partitions of nations of international system.+
3. Axelrod: (1991) "Landscape theory" a. "Distance" between invariant nations in different "partitions" of system. b. Plot mean "frustration" as surface over international configuration space. c. Stability is minimum of frustration surface. d. Alliance configuration - which nations belong to which configuration at stability.+	3. Saperstein: (1984 - 92) a. National "devotion" to armaments. b. Coupled non-linear recursion relations c. Transition from laminar to turbulent flows. d. i. Independent arms acquisition vs. balance of power acquisition and alliance jumping.* ii. Democracy vs. oligarchy* iii. Bipolar vs. multipolar*

* Achieved by changing national capabilities.

+ Pure alliance policy change; no change in national capabilities.

Table 1. "Typical" Mathematical/Physical Modeling Approaches to International Security.

of system configuration. The orbits of a dynamical model are its predicted evolution from "here" to "there"; change the model and you change its orbits and predictions. An interesting comparison of the two approaches, in the theory of games (although not further considered in this book), is the paper by Brams and Mattli (1993): "The theory of moves (TOM) brings a dynamic dimension to the classical theory of games, which its founders characterized as 'thoroughly static' (von Neumann and Morgenstern, 1944; 3rd ed., 1953: 44). By postulating that players think ahead not just to the immediate consequences of making moves, counter-moves, and so on, TOM extends strategic thinking into the more distant future. By elucidating the rational flow of moves over time... "(p.2; emphasis supplied).

In this chapter I will illustrate and compare the application of static and dynamic methods to the problem of the aggregation of larger units from collections of smaller units: how a given structure arises from its constituents. The structure might be a nation arising from its peoples, communities, or subnations (e.g., how did the thirteen states become the "United States"?) Or it might be an alliance of states such as NATO. The "problem" is used as an illustration of methods and their comparison. In the next chapter, the same two approaches will be applied to the problem of the initiation of war between these units, and a critical comparison of the two methods will be attempted. In this chapter, the goal is to predict aggregation or disaggregation (the dissolution of structure) using a recent static theory (Axelrod and Bennett, 1993) and a contemporaneous dynamic theory (Lee, Muncaster, and Zinnes, 1994). Note that the very use of the word "predict" puts these theories outside the realm of the notions of chaos and complexity introduced in the first chapter.

2. A Static Model of Aggregation

Axelrod and Bennett (1993) have created a static model which purports to explain how sixteen European nations aligned themselves on opposite sides of World War II and how nine computer companies joined competing alliances to set standards for computer operating systems. Thus, they believe the theory to be applicable to business, political, social, and security alignments in national and international systems. In addition to having the purely scientific significance of providing an understanding of how aggregation occurs in politics, economics, and society, they assume that their "landscape theory" can be a guide to policy makers. Having a coherent explanation of which international alignments are likely to form, rational policy can choose the most efficient leverage to move to the most desired configuration. Similarly, "the theory could suggest which domestic political coalitions are inherently unstable, and which would become stable if critical subgroups could be induced to make minor changes."

The theory describes the interaction of N actors, e.g., nations, each having a "size" $S_i > 0$ $(i = 1, ..N)$ reflecting its importance--measured by demographic, industrial, military, ... factors--in the system. Each pair of actors has a "propensity", p_{ij}, to work together--"a measure of how willing the two nations are to be in the same coalition together." The propensity of a pair is negative if they have, or potentially have, a conflictual relation; it is positive if there is every indication that they get along well together. The final system variable is the "distance" between the pair of actors; this number is a function, $d_{ij}(X)$, of the configuration of the system, X, how it is divided into different groupings--alliances, nations, companies,.... For example, if the problem being addressed is "which groupings will form, the distance might be taken to be 0 if both actors are in the same group, 1 if they are in different groups. Note that the $N + N^2$ numbers S and p are fixed model parameters which must be specified apriori. Similarly, though the N^2 d's are variables, their functional dependence on the system configuration X is also fixed apriori.

It is assumed that each actor will try to decrease its "distance" from those other actors for whom it has the greatest affinity = the most positive propensity; similarly it will attempt to increase its distance to those for whom it has a negative propensity. All actors will be trying to do this simultaneously and their attempts may frustrate each other. The "frustration" of a given actor i, which he will try to minimize, is defined to be

$$F_i\,(X) = \Sigma_j\, S_j\, p_{ij}\, d_{ij}(X)$$

which, for a given model (given set of parameters) depends only upon the particular system configuration--which actor belongs to which group. The importance-weighted sum of the individual frustrations, called the "energy" of the system in configuration X

$$E(X) = \Sigma_i S_i F_i (X) = \Sigma_{i,j} S_i p_{ij} d_{ij}(X) S_j$$

is a characteristic system function. Clearly, the more actors who are close to their desired partners, and/or distant from those they find repugnant, the lower will be the system energy, and conversely. Assuming that the configuration of the system can be represented as a point in a two-dimensional Cartesian Space, the system energy might be represented as a curved surface --called a "landscape"--over that flat space, as in *Figure 1.* This surface, and its higher dimension analogues, may have one or more minima, as shown in the Figure.

It is assumed that each actor will act to lower its own frustration (though no "mechanism" for so "acting" is indicated), that consequently the system will "seek" a minimum of the energy surface, that is, that individual actors will switch groups until a minimum system energy is attained. It is presumed that each switch is done by one party at a time, that the switches are sequential and that each actor knows what the system frustration is at each time and how to act accordingly. No method for carrying out these presumptions is part of the model. However, assuming a "mechanism" for moving the system over its energy surface, Axelrod and Bennett "demonstrate" that the system will--sooner or later--settle down into one of the minima of this surface, i.e., the only stable configurations of the system are at the local minima of the landscape. The metaphor is that of the system as a stone, rolling over a smooth surface of hills and dales, coming eventually to repose in the dale nearest to its starting point. Note, however, that the model itself contains no analogue of a gravitational force to impell the stone downward, nor a frictional force to hinder its motion. We just "know" that the stone will end up at a low-point; actual motion on the surface is not part of the model.

Visualizing a landscape, the region about a specific local minimum, from which a dropped freely rolling pebble would reach that minimum--low point--before coming to rest, is called the "basin of attraction" for that particular minimum. Thus a system starting anywhere in the basin of attraction of minimum A would eventually end up in the state A according to this model. An initial system configuration starting on the ridge between two basins of attraction could end up in either of two very different local minimas, depending upon an extremely small "push" one way or the other. This extreme sensitivity to an initial disturbance is called "bifurcation."

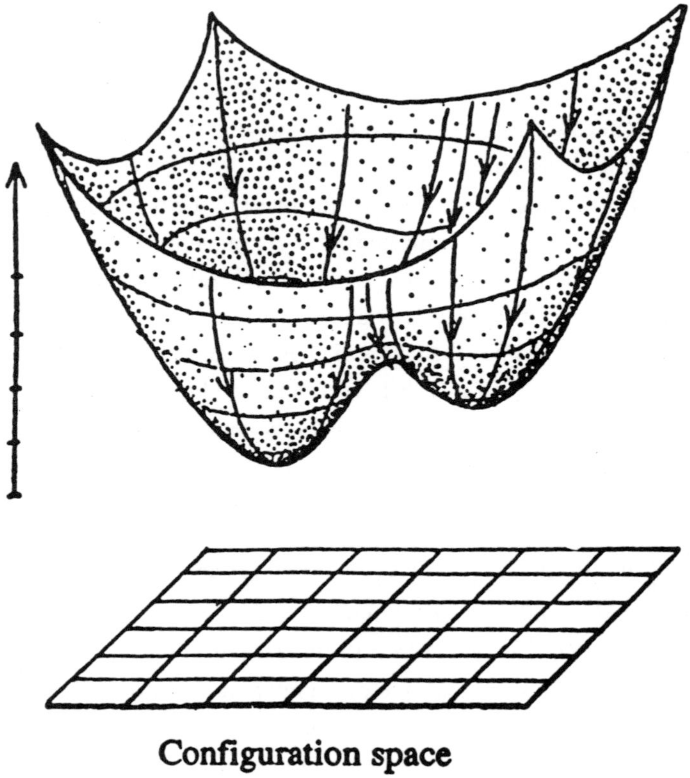

Fig.1. "Landscape" over configuration space: system "energy" versus system independent variables. (Adapted from Axelrod and Bennett, 1993).

The model has a number of interesting theoretical and policy implications, independent of a specific choice of parameters and functions.

(a) The actual world may not be the "best of all possible worlds", i.e. a given local landscape minimum may not be the lowest point in the overall landscape.

(b) The actual world configuration depends upon the history of that world, i.e., which landscape minimum it ends up in depends upon where, on the landscape, it starts.

(c) There may not be a world which satisfies everyone! No matter which landscape minimum the system has gravitated to, some or all of the actors may still be somewhat frustrated--some will find "buddies" in a different group and distasteful partners in their own group.

These model predictions agree with most "common sense" perceptions of political or human systems and thus offer little beyond a rephrasing of common experience. However, if it were possible to numerically specify a realistic landscape, i.e., have a complete set of numerical parameters which realistically described an interesting system, it would be possible to make novel quantitative predictions. For example, if the system were in one local minimum, it would be possible to specify the changes of configuration necessary (which actors should leave which group to join which other group in what sequence.) to take the system out of that minimum and head it towards another, perhaps more desirable, landscape minimum. Or, if the system were initially on a peak between two valleys, representing two different system configurations, it would be possible to ascertain which small change in configuration would lead to one valley rather than the other.

Thus, the policy maker would know what small control change to make to lead to a large change in system configuration. If we really could specify individual sizes and the propensities and distances of pairs, the final aggregation configuration of the entire system could be predicted, although we would not necessarily know how the system reached this final state.

The model was tested by attempting to "predict", starting from pre-World War II European data, which countries would be on which side in WW II, who would aggregate into the Axis Powers, who into the Allies. The nations involved, and the actual aggregations, were Britain, France, USSR, Czechoslovakia, Denmark, Greece, Poland and Yugoslavia versus Germany, Italy, Hungary, Estonia, Finland, Latvia, Lithuania, and Rumania. The size parameter of each of these sixteen nations was determined using only the six components of demographic, industrial and military power, as measured with the national capabilities index of the Correlates of War project (Singer, Bremer, and Stuckey, 1972). The propensity factors were estimated using information, for each actor-pair, on the presence of ethnic conflict, religious similarity of populations, border disagreements, history of wars between them, and similarity of governments. It was assumed apriori that there would only be two alliances, but each of the sixteen states could belong to either; thus there were $2^{16}/2$ = 32,768 possible configurations in 1936. The resultant landscape of 32,768 points, found by Axelrod and Bennett, had two local minima. The one with the larger basin of attraction, whose minimum was also the global minimum, corresponded to the exact actual alignment of European states in World War II with the exception that it erroneously placed Poland on the German side. The smaller basin was characterized as a "pro-vs.-anti Soviet alignment." Thus, three years before the outbreak of the war, the landscape model predicted the alignment of WW II "quite well". After testing other static models, Axelrod and Bennett conclude "that in a static analysis, landscape theory is superior to alternative methods of computing likely adjustments." They repeat the exercise for 1937, 1938, and 1939, changing the model parameters (the national capabilities and propensities) appropriately. By 1939 they find only one local minimum, the correct one, with Poland now correctly aligned against Germany.

This result, and further similar studies of post-World War II European alignment, as well as of corporate alignments in the computer industry, gives this model a great deal of apparent credibility. However, it should be emphasized that the model is purely static: it provides no mechanism ("explanation") for how the system moves from one alignment to another in a given basin of attraction. Furthermore, one would expect that the system parameters would change as the system configuration changes, e.g., Germany's "relative size" is increasing throughout the decade of the '30s, because of its rapidly growing military expenditures, due, presumably, to the rapid approach of the system to its final war time alignment. Axelrod and Bennett had to put these changing parameters (their "mobilization for war") in "by hand"--from outside of the model. A more complete description of conflict-alignment should certainly include the evolution of capabilities with evolving configuration.

3. A Dynamic Theory of Aggregation

A dynamical theory of aggregation attempts to explain the process whereby a system gets from one configuration to another. It thus describes the rate of change of the system in a configuration space, e.g. "motion" over a landscape surface such as that of the previous section. Hence we expect its basic elements to be "velocities": time-derivatives of variables represented the alignments of interest, e.g., the "distances" between states in the system of interest. These *rates* are to be related, via the model, to the variables appropriate for describing the system so that the system changes its alignment because of its alignments and the properties of its constituent states. One such theory is that of Lee, Muncaster, and Zinnis (LMZ, 1994). The question they attempt to answer is whether states in the international systems will come together in alliances (or even into multi-state nations or whether they will drift apart into antagonistic or even hostile relations.

The system variables chosen by LMZ to describe an N nation system are r_{ij} (i,j = 1,...,N, i≠j), the N(N-1)/2 "relationships" between nations i and j. Relationship is assumed characteristic of the pair (i,j), not of one for the other; thus $r_{ij}=r_{ij}$. Positive values for h(r_{ij} = 1, ...,N, i ≠j), the N(N -1) "relationships" between nations i and j. Relationship is assumed characteristic of the pair (i,j), not of one for the other; thus $r_{ij} = r_{ji}$. Positive values for the relationship imply friendship and/or cooperation between the pair; the greater the value, the more tightly the ties between the pair. Negative values imply hostility and eventually, if large enough in magnitude, war. The relationship r_{ij} is a measure of intent and capability of i and j towards each other. Thus, for example, in a three nation system, if r_{12} and r_{13} are negative whereas r_{23} is positive, the implication is that parties 2 and 3 are aligned with each other against party 1. The problem then is, given an initial set of relationships (the values of all r_{ij} at some starting time), to predict the eventual alignments of the system. The rates of concern are the variations, with time, of the r_{ij} , usually designated as "$\dot{r}_{ij}$ ", the time derivatives (calculus!) of the relationships = dr_{ij}/dt. If these $\dot{r}_{ij}$ can be uniquely related to the values of the r_{ij}, the result will be a set of differential equations. Solving these, given the initial set of relationships, should produce unique predictions of the evolution, with time, of the r_{ij} and thus of the eventual alignments of the system.

The model chosen by LMZ to relate the $\dot{r}_{ij}$ to the r_{ij} is the mathematical formulation of the set of rules:

The friend of my friend is my friend, (1)

the friend of my enemy is my enemy, (2)

the enemy of my enemy is my friend, (3)

the enemy of my friend is my enemy. (4)

Presumably this set of rules describes the relationships between nations, individual people, and all levels of social organization between the two extreme levels. Rule (1) can be translated as $r_{ij} > 0$ and $r_{jk} > 0$ implies $r_{ik} > 0$; if i and j are friends, and j and k are friends, then i and k must be friends. The translation of rule (2) is: $r_{ij} < 0$ and $r_{jk} > 0$ implies $r_{ik} < 0$; if i and j are mutually hostile, whereas j and k are friends, then i and k must be hostile towards each other. Rule (3) is expressed as: $r_{ij} < 0$ and $r_{jk} < 0$ implies $r_{ik} > 0$; if i and j are hostile to each other and j and k are also a hostile pair, then i and k must be a friendly pair. Finally, rule (4) says: $r_{ij} > 0$ and $r_{jk} < 0$ implies $r_{ik} < 0$; if i and j are friendly, and j and k are a hostile pair, then i and k must be a hostile pair. All four of these rules are contained in the LMZ postulate that r_{ik} be proportional to the product of r_{ij} and r_{jk}:

$$r_{ik} \sim r_{ij} \bullet r_{jk} \tag{5}$$

The rules are turned into a dynamical model by expressing them as rates, i.e., if $r_{ij} > 0$ and $r_{jk} > 0$, or if $r_{ij} < 0$ and $r_{jk} < 0$, then r_{ik} is increasing; the relationship between i and k grows warmer and closer if both are friends of j or if both are enemies of j. Similarly, if $r_{ij} > 0$ and $r_{jk} < 0$, or if $r_{ij} < 0$ and $r_{jk} > 0$, then r_{ik} must be decreasing; the relationship between i and k becomes colder and more distant if i is friendly towards an enemy of k or if i is an enemy of a friend of k. These revised rules are incorporated in the revised postulate:

$$\begin{aligned} &r_{ij} \bullet r_{jk} > 0 \text{ implies } r_{ik} \text{ is increasing: } \dot{r}_{ik} > 0 \\ &r_{ij} \bullet r_{jk} < 0 \text{ implies } r_{ik} \text{ is decreasing: } \dot{r}_{ik} < 0 \end{aligned} \tag{6}$$

The <u>simplest</u> mathematical relation incorporating the postulate (6) is

$$\dot{r}_{ik} = B_{ikj}\, r_{ij}\, r_{jk} \tag{7}$$

where B_{ikj} is a positive constant, characteristic of the system. So far, the rules have been stated as if there were only three nations in the system: i, j, and k. If there are N>3 nations, LMZ simply assume that any single nation, i, interacts with each pair, j and k, in the system according to the above rule and that the results are cumulative, i.e., just add the time rates of change for each pair to get the total rate:

$$\dot{r}_{ik} = \Sigma_j B_{ikj} r_{ij} r_{jk} \tag{8}$$

where the sum is over every j (1,...,N) not equal to i or k and the result is a set of N(N-1)/2 coupled equations for the N(N-1)/2 variables r_{ij}, with N(N-1) (N-2)/8 parameters B_{ikj}.

The mathematical rules captured by the equations of (8) imply that the interactive behavior of the pair (i,k) is governed solely by the relations with all other possible pairs. But the existing relationship within a pair must have some impact on the future evolution of that pair. LMZ assume that this impact is one of endurance: positive relations lead to an increasing relationship whereas negative relations lead to a decreasing (increasing hostility) relationship. This assumption is expressed most simply as

$$\dot{r}_{ik} = A_{ik} r_{ik} \tag{9}$$

where A_{ik} is a set of N(N-1)/2 positive system parameters. The complete model, combining Eqs.(8) and (9), is then

$$\dot{r}_{ik} = A_{ik} r_{ik} + \Sigma_j B_{ikj} r_{ij} r_{jk} \tag{10}$$

The set of coupled, non-linear, first-order differential equations (10) has a unique solution, $r_{ik}(t)$, given a unique set of system parameters and starting values $r_{ik}(0)$. Given a set of these constants, the solutions could be obtained numerically on a computer, thus giving an exact evolution of the system alignment with time. However, in general these constants are not going to be known precisely, or even approximately. Even if approximately known, the non-linear equations may be so sensitive to small changes in the input parameters that the output solutions will be meaningless. (This sort of sensitivity will be extensively discussed in Chapter 4.) Hence LMZ sought general theorems which would be applicable no matter what values were specified for the system parameters. The hope was that such general theorems would provide some useful general insight into the evolution of the real international system.

They start by defining two specific alignments of nations: "unipolar" and "bipolar". A unipolar configuration is one in which all relationships are friendly: all r_{ij} are positive. A bipolar configuration is one in which the N nations are broken into two groups A and B; all nations within A have friendly relationships with each other; similarly with the nations in B; but all nations of A have hostile relationships with those of B: all $r_{ij} < 0$ if i is in A and j is in B. Given these definitions, they are able to prove two general theorems:

Unipolarity Theorem: A system which is initially unipolar will remain unipolar for all time. Furthermore, the bonds between the nations will grow ever more positive, exponentially, in time: all r_{ij} will become infinitely large and positive with the indefinite increase in time. If one postulates that an ever increasingly positive tie between two nations is to be interpreted as an eventual amalgamation of the two nations into one country, then a bunch of friendly countries will eventually become integrated into just one country.

Bipolarity Theorem: A system which is initially bipolar, with configuration (A,B) will remain bipolar with the same configuration: All pair-bonds within A, and all bonds within B, will grow increasingly positive, exponentially. All pair-bonds between a nation in A and a nation in B will grow increasingly negative, exponentially in time. Thus, given the postulate immediately above, the two groups of friendly nations will eventually become two nations. And, if we postulate--as seems reasonable--that an increasingly hostile relationship will sooner or later lead to war, then the theorem implies that war between the two eventual nations is inevitable.

If the system is initially neither unipolar nor bipolar, conclusions are more difficult to obtain. LMZ restrict themselves to N=3 "homogeneous" systems: all the A parameters have the same value, as do all B's. Here they were able to show that "almost all systems that are not initially unipolar or bipolar become so after a period of time." From what has been said before, almost all such systems will either end up as a single nation or as a system of two nations at war with each other.

The predictions of this simple model are admittedly dismal. The model of course is overly simple and leaves out many factors presumably important in the evolution of domestic and international affairs. (For example, it is a model only of intents--simple at that, omitting all constraints of capability. Nations usually try to avoid war if they possess no capability to wage it; nations with overwhelmingly large capabilities usually do not have to wage war.) However, the model does illustrate the essence of a dynamical theory: the state of the system at any one instant of time determines the state of the system at the immediately following instant, i.e., the rate of change of the system's configuration is determined by the system itself. There is no need to go outside of the system and its rules to obtain additional parameters or rules for change, as is the case with a static theory.

4. References and Additional Readings

Axelrod, Robert and Scott Bennett. 1993. "A Landscape Theory of Aggregation", British Journal of Political Science, **23**, 211-33, April.

Brams, S. and W. Mattli 1993. "Theory of Moves: Overview and Examples", Conflict Management and Peace Science, 12, No. 2. 1-40.

Intriligator, Michael D. and Dagobert L. Brito. 1984. "Can Arms Races Lead to the Outbreak of War?" Journal of Conflict Resolution 28: 63-84.

Kaye, J.D. and G.R. Lindsey. 1970. "MIRVs and the Strategic Balance", Nature **227** : 696-7.

Lee, S.C., R.G. Muncaster, and D.A. Zinnes 1994. "The Friend of My Enemy is MY Enemy: Modeling Triadic Internation Relationships",Synthese **100**: 333-358.

Richardson, L.F. 1960a. Arms and Insecurity, Pittsburg: Boxwood Press.

Richardson, L.F. 1960b. Statistics of Deadly Quarrels, Pittsburg: Boxwood Press.

Saperstein, Alvin M. 1990. "Chaos and the Making of International Security Policy," in The Ubiquity of Chaos, edited by Saul Krasner, Washington, D.C.:American Association for the Advancement of Science.

Saperstein, Alvin M. 1991. "The Long Peace - Result of a Bi-Polar Competitive World?", J. of Conflict Resolution, 35 (No. 1): 68-79.

Saperstein, Alvin M. 1992a. "Are Democracies More or Less Prone to War? A Dynamical Model Approach", Mathematical and Computer Modelling, 16 (No. 8/9): 213-221.

Saperstein, Alvin M. 1992b. "Alliance Building vs. Independent Action: A Non-Linear Modeling Approach to Comparative International Stability," J. of Conflict Resolution, 36 (No. 3): 518-545.

Singer, J. David, Stuart Bremer, and John Stuckey. 1972. "Capability Distribution, Uncertainty and Major Power War, 1920-1965." in Bruce Russet, ed. *Peace, War, and Numbers.* Beverly Hills: Sage.

von Neuman, John and Oskar Morgenstern. 1953. "Theory of Games and Economic Behavior", Third Edition. Princeton, NJ: Princeton University Press.

CHAPTER 3
PREDICTIVE MODELING OF ARMS RACES: STATIC AND DYNAMIC

The previous chapter was devoted to the predictive study of structure building, using static and dynamic models. We now turn to predictive models of war initiation, again contrasting one illustrative static approach with one dynamic model. War initiation may be taken to be the opposite of structure building in that the potential destructiveness of modern nuclear war implies the dissolution of natural structures, perhaps of the entire international system. This chapter then concludes by comparing static and dynamic predictive models and discussing the shortcomings of both.

1. A Static Model of War Initiation

During the height of the nuclear confrontation between the Soviet Union and the United States, great fear was expressed on both sides about the possibility of a sudden initiation of full scale nuclear war. Presumably one side would attempt to disable the other by launching a major nuclear missile strike - either "out of the blue" (with no readily apparent build up of hostilities) or more likely, as a consequence of a non-nuclear military confrontation which escalated to a strategic nuclear exchange. During that time of intense confrontation, and fear of its consequences, a number of scholars introduced static models of the hostile binary world system. The hope was that these models would capture enough of the reality of the situation so that their analysis would enable rational policy makers to avoid the nuclear holocaust which many feared to be the likely consequence of the intense hostile competition between the two sides. Though they differed in details, these models (e.g., Kay and Lindsey, 1970; Legault and Lindsey, 1976; Intrilligator and Brito, 1984) shared a basic static approach.

They assumed the existence of two hostile competing nations - no attempt was made, within the models, to explain the origins or continuation of the hostility, nor how the world had evolved into a basically bipolar system. The fundamental variable of each of these models was a quantitative measure of the major threat which each side represented to the other. In the latter half of the twentieth century, the prime threat to the existence of a major nation was the delivery of nuclear weapons at long range by missiles and/or aircraft. Hence, the variable representing each side was a numerical description of the nuclear explosives it could deliver to the other side by bombers, sea-based, and land-based missiles (ballistic and cruise). Complex calculations might be needed to obtain these numbers since they could change as a result of being successfully targeted by the opponent's nuclear weapons- whose numbers would change as the weapons were used.

If side X's number, M_x, were sufficiently greater than the corresponding number for side Y, M_y, then under ideal conditions, X could launch a first strike which would knock out all of Y's weapons, thus depriving Y of the possibility of a retaliatory strike. X would still have enough weapons left over so as to threaten the cities, population, and resources of Y with utter destruction. Thus, the disarmed Y would be compelled to bow to the wishes of X, since X could commit "murder" to enforce its will. Y is unable to deter X from "pushing it around" for, if Y strikes first, X has a sufficient number of weapons to absorb all of Y's first strike and, still have enough surviving weapons to be able to then severely punish Y in retaliation.

Since X can "commit murder," and get away with it, in support of its goals, it has no need to commit the murder. Furthermore, there would be no gain from such a murder; the devastated victim nation can't give "victorious" nation X what it desires from Y. Similarly, there is no incentive for Y to strike first since both parties know that X has a guaranteed devastating retaliation due to X's initial weapons excess over Y. Hence, neither side has an incentive to respond provocatively to slight political, economic, or military insults from the other. Furthermore, if Y attempted to increase its forces so as to eliminate X's dominating surplus, X could increase its own weapons stocks correspondingly - assuming no information, economic, resource, or political constraints - to maintain its superiority. Hence, there is no incentive for Y to attempt to change the situation; and there certainly is no incentive for X to change. The resulting situation is "stable" - small changes to the situation from either party or from outside of the system, require no major response from either side.

This stability is desirable if one's sole goal is the avoidance of war. It is not particularly desirable from the point of the dominated nation Y. Of course, if all the numbers were exchanged, $M_x \leftrightarrow M_y$, the roles of X and Y would be reversed. The relation between X and Y is perfectly symmetric; neither side has an inherent advantage distinct from the numbers under consideration. It's the discrepancy between M_x and M_y which produces the asymmetry in dominance. The actual amount by which M_x must exceed M_y in order for X to stably dominate Y depends upon the details of the model: the assumed ability and accuracy of the missiles, the number of warheads per missile, the number of attacking and attacked missiles, the efficiency of each party's C^3I (Command, Control, Communications, and Intelligence). Working out these details and plotting M_X and M_y against each other as a Cartesian plot (Fig. 1), which I will refer to as a "deterrence plot," the result is a graph with regions of stability. If the initial configuration of the bi-national system lies in one of these stability regions, the system will remain stable, assuming "rational policy making" and barring accidents.

Alternatively, if the two sides are evenly matched, $M_X \sim M_y$, the situation is again stable. The first striker, launching all of its weapons, can't wipe out all of its victim's weapons, leaving itself open to a destructive retribution. Such a murderous first strike thus leads to the suicide of its perpetrator. Presumably neither side wishes such a murder-suicide , meaning that again neither side will

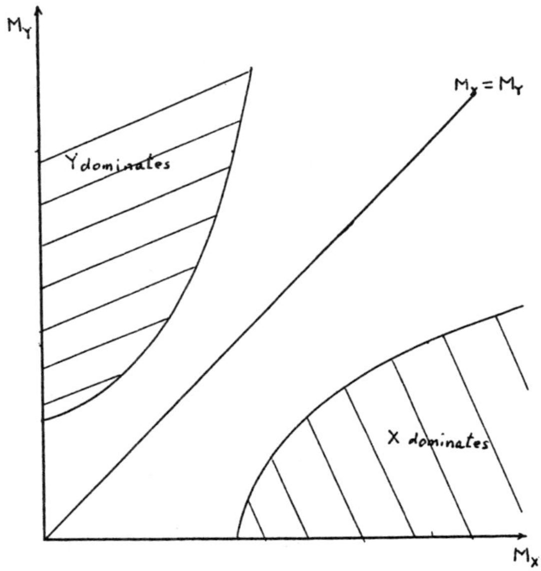

Fig.1 Deterrence Plot: numbers of missiles, M_x versus M_y, of hostile pair of nations, X and Y, threatening each other.

make a major riposte to a minor slight--which is the definition of crisis stability. So this equal strength configuration is represented by another stability region on the M_X vs M_y plot, though this time the stability is symmetric in that neither side can dominate the other.

In between these stability regions--areas on the M_X versus M_y plot in which there is no gain to a first striker - there are regions where there is an advantage to the side which strikes first. For example, the numerical difference between M_X and M_y may allow X to infer that if Y strikes first, X will be unable to retaliate, whereas if X strikes first, Y will be unable to retaliate. Thus, whenever the international situation gets tense enough for X to believe that Y may lash out, X will be under strong pressure to launch a first strike. Hence, a minor altercation between X and Y may lead to catastrophic results, a full nuclear exchange. Such a region in the M_X - M_y plot would be called "unstable" and is the graphical description of the situation labeled "crisis unstable" by the political scientists. It represents an international configuration which is to be strenuously avoided by those who believe that even a "successful" first strike is a major human defeat.

A more realistic deterrence plot would be like Fig. 2 (suggested by the calculations of Legault and Lindsey, 1976). Here, stability may be due to the weapons dominance of one side or to the lack of weapons by both sides. When both sides are roughly equivalent there is stability when weapons stocks are large; neither can assume that a surprise initial attack will wipe out all of the victims missiles, thus preventing a devastating retaliation. However, with equal intermediate levels of missiles, the first one to strike may be assured of no retaliation; the laggard is assured of devastation and defeat. There is thus great incentive to strike first whenever hostilities may seem immanent. This very dangerous region is "doubly unstable".

Of course, the discussion so far is based upon complete rationality - of humans, their facilities and equipment, and of nature. It assumed that there is no possibility of a human, electronic, or mechanical failure which will cause one side to launch a nuclear attack even if the world is sitting in a stable part of the deterrence plot. We know that such accidents can occur - almost disastrous "accidents" have occurred - and that, everything else being equal, the probability and consequences of such an " accident" increase with the number of weapons available. Hence, everybody's security would be enhanced, and costs considerably lessened, if the world could position itself near the origin of the deterrence plot: $M_X \cong M_y \sim O$. We are not in such a region now.

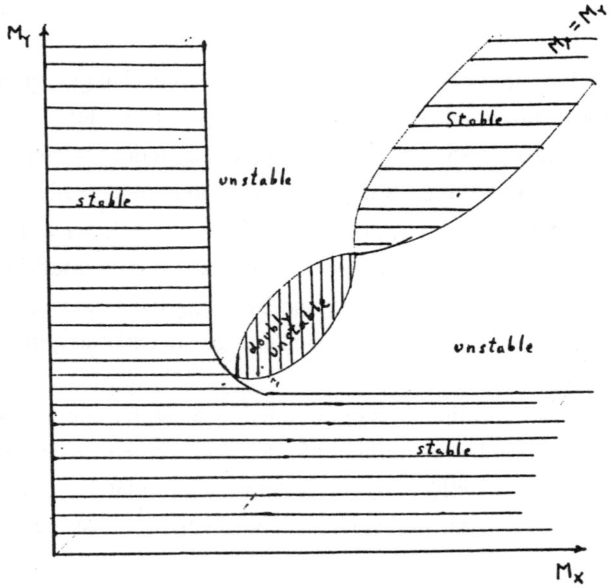

Fig.2. A more realistic Deterrence Plot.

So, a fundamental question facing the policy makers of the major world powers is how to get from the present world nuclear-weapon configuration-- whether unstable or stable, but with dangerously large numbers of accident-prone weapons -- to a more desirable stable, less expensive, more secure region. If the assumptions underlying the deterrence plot are reasonable, the static model shows policy makers which regions of nuclear-weapon configuration space are desirable, which are undesirable. But it gives no indication as to how to move in this configuration space from one region to another, how M_X responds to changes in M_y and vise-versa. By the very definition of "static model", there are no "forces" in the model which can bring about change.

Without knowledge, and control, of such forces, how can we safely move from one stable region to another, more desirable, stable region? How do we get from the present "here" to the future "there?" If an unstable region in between must be traversed, how can this be done in the least dangerous manner? Static models offer few clues to aid the policy maker dealing with such fundamental world problems.

2. A Dynamic Model of War Initiation

A popular model for the response of one nation to the changes in military capability of its competitive partner in the hostile world system is due to Richardson (1960 a,b; Hess, 1995). Again, nothing in the model indicates why or how the system became competitive. It is just assumed that, in a "realistic world", nations are inherently suspicious of each other (c.f. Kegley, 1993, for a discussion and references on "structural realism") If one party builds up its military strength, others in in the system will immediately respond by doing the same! That is, of course, assuming that they are capable of doing so, given existing political and resource constraints.

The military capability of nation X is represented by a single, one-dimensional, numerical variable x; x might be the number of men (and women) under arms, the sum of the numbers of all available weapons systems- appropriately weighted, the number of nuclear weapons (and, perhaps, other weapons of mass destruction - chemical, biological, . . .) ready for use, the total budget (in readily comparable monetary terms) devoted to military purposes, or the fraction of gross national product invested in weapons systems. Similarly, the military capability of nation Y is the positive real number y. The time rate of change of these variables - the rate at which these weapons and/or people are procured or decommissioned, their derivatives with respect to time, are designated $\dot{x}$ and $\dot{y}$. The basic hypothesis of Richardson, and his followers, is that these rates are linearly related to the variables themselves, that $\dot{x}$ and $\dot{y}$ are proportional to x and y.

It is assumed that the greater the military capability of Y, the greater the threat it represents to X, and hence, the greater the rate at which X will increase its own military capability, and visa-versa. Thus, $\dot{x} = ay$ and $\dot{y} = bx$, where a and b are positive parameters representing the innate ability of the two nations to acquire arms and their hostility towards each other. However, the more military capability already possessed by X, the less it needs to acquire additional capability - and the less its population will be willing to make the required economic sacrifices. Hence, $\dot{x} = -cx$ and $\dot{y} = -dy$, where c and d are positive real parameters representing the military "complacency" and resistance to further military expenditures of the two countries, (Of course, it is possible that the existence of a large military establishment - a "military - industrial - political complex," or an "iron-triangle" - might produce the effective political pressures necessary to further increase those military capabilities. In that case, c and/or d would become negative real numbers). Finally it may be that nation X and/or Y would acquire arms independently of the arms stocks of its opponent/ partner, that arms capability would be acquired - or diminished - on the basis of the perceived state of relations, hostile or friendly, between the two states. In this case, $\dot{x} = e$, $\dot{y} = f$, where e and f are real numbers representing that perceived relationship; they would be positive numbers if the perceived intent of the partner is hostile, requiring an increase in military capability, negative if the relationship is friendly, allowing a decrease in armaments.

Putting these three linear assumptions together, the standard Richardson equations, purporting to describe the competitive security interactions of a pair of nations, are obtained:

$$\begin{aligned} \dot{x} &= ay - cx + e \\ \dot{y} &= bx - dy + f \end{aligned} \tag{1}$$

(It is easy to extend this model to a group of N > 2 nations, e.g., see Saperstein, 1994.) Constants a and b are measures of the response of each nation to the "capability" of its opponent, c and d measure the response to one's own "capability" whereas, e and f are measures of response to the <u>intent</u> of the other nations. Thus a and b are measures of the distrust and suspicion felt for the opposing nations, whereas, c and d measure the complacency with one's ability to deal with possible threats from the outside.

The question naturally arises: how can the response of a nation to capability and/or intent be represented by a single constant number? The possibility of representing capability by a single number, x or y, is already presumptuous - accounting, perhaps well, for the quantity of weapons and personnel, but with more doubtful efficacy for the quality of weapons systems and human leadership, training, willpower, etc. (This numerical presentation of military capability is similar to an assumption used in the previous section on static models). But the <u>response</u> of a nation is a much diffuse concept; it is the manifestation of a shifting conglomerate of economic, political, cultural, historical, sociological, geographical, etc., factors working through, influencing and being influenced by, many individuals (in a modern democracy- many millions of individuals). How can all of this be reduced to a single number, per nation; independent of time?

As indicated in the introductory chapter, these national-behavior-describing parameters are presumed to be averages over the appropriate behaviors of the many individuals making up the nation. Their time variation is that of the evolution of *individual and cultural behavior patterns* which, as indicated by common observation, marches to a much slower beat than do *national security policies.* They can be influenced by policy leaders, but change occurs slowly. Hence, we assume the former to be effectively constant when considering the latter. In an ideal, reductionist science, the values of these parameters would be deducible from the known behaviors of individuals. In the practical science which we are attempting to create, their values will have to be obtained by observation of actual national behavior. The assumption is heuristic: its validity is to be judged by the effectiveness of its application--its success in furthering understanding of the international security system and facilitating the creation of useful policy for dealing with that system.

Given the parameters and the initial values of the variables--the values of x and y at some given starting time, x_0 and y_0--the Richardson Equations can be uniquely solved for x and y as functions of time, x(t) and y(t). The future military capabilities of the two nations are completely determined and can be represented by a unique curve--a "trajectory" in the two-dimensional x-y space. If a nation's leaders are not satisfied with their predicted future, they might try to change the outcome by changing one or both of the initial military capabilities, x_0, y_0. It would be much more difficult to change the system parameters, a, b, ..., f since modifying them would involve mass changing of individuals, their culture, and their society.

Richardson was not interested in the specific evolutions of military capabilities in the world system--which, of course, were sensitive to the specific starting values--which were subject to fluctuation. He wanted to know if the system was necessarily going to fall into war or if it were possible that there are sets of parameters (national behaviors) which would guarantee peace. In order to deal with this question, he had to define peace and war, which, in this case, he took to mean stability and instability. A stable situation, according to Richardson, was one in which the military capabilities of the system would not grow with the passage of time--would necessarily tend to finite constants with the increase of time, no matter how the system started. Conversely, an unstable situation was one in which the military capabilities of one or more of the nations in the system would necessarily grow beyond all bounds with the sufficient elapse of time. Of course, unlimited growth of military capability (or any other capability) is impossible in a real world; resource limitations--not included in the Richardson Equation approximation to the real world--would make their presence felt sooner or later.

Even when such limitations are not expected to be significant--when the resources devoted to military capability are a vary small fraction of the total resources easily available to the nation in question--there has been much debate among political scientists (e.g., Wallace, 1979) as to whether or not an ever increasing military capability necessary implies impending war or, conversely, whether steady or decreasing military capability really implies a peaceful era. Nevertheless, Richardson sought the conditions which would guarantee his definition of stability or instability.

The most general solution to the Richardson Differential Equations, Eq. (1), can be written as

$$x = \alpha + (\beta + \gamma t)\, e^{\lambda t}$$
$$y = \xi + (\delta + \eta t)\, e^{\lambda t} \qquad (2)$$

where the real constants α, β, γ, ξ, δ, η, and the complex constant λ (Greek letters: alpha, beta, gamma, tsi, delta, eta, lambda) are determined by the system parameters a, b...f, and the initial values x_o and y_o of the system's military capabilities. If the real part of λ, λ_R, is a positive number, the solutions, (2), grow exponentially with time t. They are unbounded, hence unstable in Richardson's view. A negative real part means that the time varying part of the solution exponentially decreases to zero as time increases positively from the initial time. In this case the solution is Richardson stable. If the exponent parameter λ has an imaginary part, λ_I, this just means that the solutions have an oscillatory motion (as sin $\lambda_I t$) superimposed upon the exponential growth or decay. It is the latter which determines stability or instability.

Substituting the assumed solution (2) into the system-determining-equation (1), it is an elementary algebraic exercise to show that a solution is only possible if λ satisfies the quadratic equation

$$\lambda^2 + (c + d)\,\lambda + cd - ab = 0 \qquad (3)$$

which has two solutions, λ_1 and λ_2:

$$\lambda_{1,2} = (1/2)\,\{ -(c + d) \pm [\,(c + d)^2 - 4(cd - ab)]^{1/2}\} \qquad (4)$$

Thus, if cd - ab < 0, there are two real solution, one positive, the other negative: $\lambda_1 > 0$, $\lambda_2 < 0$. Since the actual solution, for a given set of starting values, must contain a sum of two terms like (2), one with λ_1 in the exponent, the other with λ_2, it follows that the full solution must be exponentially increasing: it is Richardson unstable. If cd - ab = 0, $\lambda_1 = 0$ and λ_2 is negative-real; there is no exponential growth. If $(c+d)^2 / 4 > cd - ab > 0$, both roots are real and negative; the exponential terms decay to zero leaving only constant military capabilities. Finally, if $cd - ab > (c + d)^2 / 4$, the two roots are complex, with the same negative real part, and imaginary parts equal in magnitude but opposite in sign; the exponential terms oscillate as they decay to zero. (see Fig. 3)

Thus the future of the Richardsonian system depends upon the quantity cd - ab, the difference between the product of the "complacency" factors and the product of the "suspicion" factors. If the "net complacency" is greater than the "net suspicion", the system is stable; otherwise, it is unstable. Note that the predicted stability or instability of the world system, according to Richardson, does not depend upon the intent of the participating nations towards each other--the parameters e and f. Stability is driven exclusively by reactions to capability, not to intent. This seems contrary to the usual expectations but is just the kind of result Richardson was seeking: war and peace are determined by the military capabilities of the systems and reactions thereto, and are not due to human feelings of friendship or hostility. The policy-maker seeking peace must increase the "complacency", dampen the "suspicion" of the system--perhaps easier said than done.

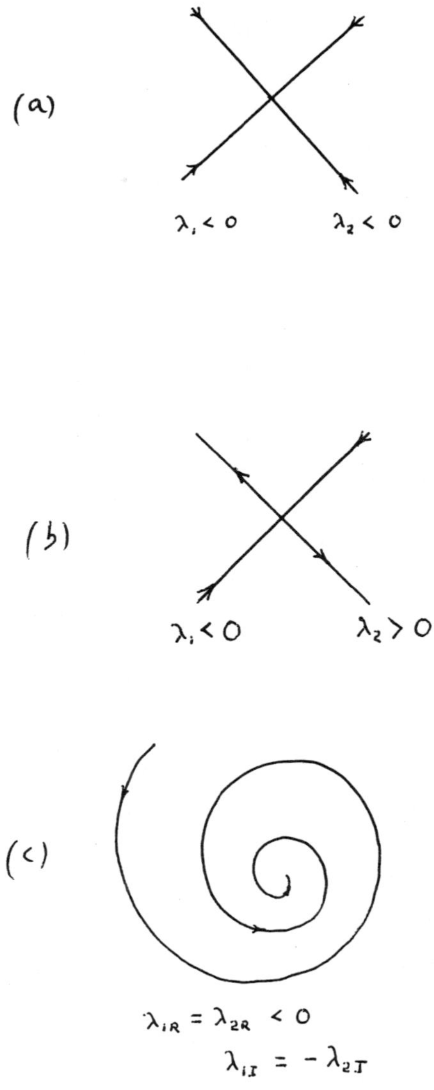

Fig. 3. Approach of Richardson Model Solutions to their asymptotic value in "phase space" (x versus y). (a) Both roots real and positive; hence both linearly independent solutions decay exponentially to the limit point. (b) Both roots real, one negative, one positive; one solution decays exponentially to limit point while the other grows exponentially away from it. (c) Two complex conjugate roots with negative real parts; both solutions oscillate as they exponentially decrease towards the limit point. If the real parts were positive, both solutions would grow exponentially; there would be no limit point.

3. Problems of Static and Dynamic Models

Two types of static models have been examined. In one, stability is determined as the minimum (or maximum) of some sort of surface in a configuration space whose dimensions represent the security characteristics of the international system. In the second, stability lies in regions of such a space whereas other regions represent unstable security configurations. In both types, time is absent from the model and hence the theory is unable to describe how the system evolves from one point in the space to another, moving from stable to unstable configurations or vice-versa. Lacking time, static theories are fundamentally incomplete. Even if they included time--if they were dynamic rather than static--models and theories of international competition and security are necessarily incomplete: only the actual system can be an absolutely accurate model of itself.

The incompleteness of a static maxima/minima theory implies that stabilities indicated by the theory are likely to actually be unstable. To increase the correspondence between any theory and the reality it is supposed to represent, additional variables (dimensions) and parameters are required. It is very easy to show, geometrically, that a maximum or a minimum in a function of n independent variables is unlikely to remain a maximum or a minimum as additional variables come into play. An n-fold maximum or minimum point of surface may become a saddle-point of an n+1-fold surface or, more likely, a completely undistinguished point. (The essence of a "saddle-point" is that it is a minimum along one curve in the surface and a maximum for the curve, in the surface, traversing that point perpendicularly to the first curve.) Thus, for example, the minima of the two- country energy surface of Fig. 1 in Chapter 2, which are presumed to determine the stable configurations of two nations, may lose their minimum status when a third nation becomes significant in the international system. If the presence of the third nation turns a minimum into a saddle-point, the prediction of a stable equilibrium configuration itself becomes unstable and untrustworthy, especially since nothing in the model indicates which path the system might take over the saddle. Hence, would-be-users of such models must have considerable insights into the effects of presumably minor subsidiary variables. But the whole point of modeling a system--in the physical or social sciences--is to minimize the number of variables which require concern. Thus it seems unlikely that much scholarly insight will result from such models.

A more fundamental criticism arising from the incompleteness, especially for the policy makers, is that the process of going from one model configuration to another may significantly change the system being modeled. The process of changing configurations is interactive but this interaction may not be included in the model. This difficulty is most easily illustrated via the 2-nation stability plot of this chapter: Fig. 2. In order to move from a given "dangerous" initial region of the plot to a more desirable configuration, at least one of the nations is going to have to engage in a weapons build-up or build-down. The other nation seeing just the change in weapons configuration of its opponent, and being ignorant of the motivation (the usual situation!), may respond by matching or overmatching a build-up or by attempting to "take advantage" of the presumed weakening indicated by a build-down. For example (Fig. 2), suppose the dangerous starting point is within the lower left end of the "doubly unstable" region. X, recognizing its danger, builds up its missile stocks in the hope of reaching the "X not deterred" region. Y acts analogously. The result is that the final configuration is still in the original danger zone. More importantly, and not included in the model at all, is the perception by each party that the other, hostile party, is striving for an advantage. This perception will certainly alter, in a dangerous direction, the estimates upon which the model is based. The attempt by part Y to change its position on the static stability plot may provoke the other party into a preemptive attack, preemption either to prevent the "weaker" opponent from becoming strong enough to attack the currently stronger, or to blunt the expected attack of a stronger opponent and hence minimize the damage to oneself (one possible interpretation of the Japanese "surprise" attack on Pearl Harbor, initiating the U.S. participation in WWII). The point is that none of these possible dynamic responses are included in, or suggested by, the static model which is thus, of itself, incompetent to describe and predict the process and its outcome. Clearly, wise policy making requires more than the ability to say that one's present configuration is or is not stable. You must be able to predict how the system will evolve from its present configuration, given the policy choices you are contemplating. You need dynamics!

The dynamical evolution of the system required for rational policy making is most simply viewed via Fig. 1a in chapter 1. The present configuration of the system is the starting end of the "trajectory" (also called "orbit") which represents the unique evolution of the system into the future. The orbit is the unique result of the dynamical equations determining the system and its starting values. If the predicted configuration of the system at some future time (the value of the orbit at that time) is unsatisfactory, the policy maker must either change the starting values or the dynamical laws governing the system. However, the determination of the actual starting values of the system can never be done precisely. Due to unavoidable errors of observation, the starting values are only known within some range. Hence, prediction should really be represented by Fig. 1b (a group of orbits starting within the initial range)), rather than 1a (a single orbit from a single starting point), of Chapter 1. As long as the two orbits, representing the range of the starting values, remain close to each other, uncertainty as to the future state of the system is roughly the same as the measurement uncertainty of the present. Hence the single orbit prediction of 1a is useful. But many dynamical models are "non-linear", as will be discussed in the next chapter, which implies that some initial configurations are dynamically unstable (e.g. Schuster, 1988)--initially neighboring orbits may separate exponentially as in Fig. 1c of Chapter 1. Such instability may make prediction impossible from these initial configurations since a "spread" in initial values is always possible and there is usually "system noise" along the orbit.

The role of observational error in spreading out the possible starting configurations of a system is well known but something more should be said about the random disturbances--"noise"-which occur everywhere along a system orbit. Any finite model of a real political system is necessarily incomplete. A nation is not one or a few variables. It is a collection of millions of people, each of whom has a complex psychology of millions of possible actions and interactions. The relatively few variables used to describe such a system, in any potentially useful model, will each be averages over the behaviors of countless individual behaviors (just as the pressure of a gas is an average description of the behavior of its effectively countless molecules). There will always be random fluctuations about these gross average variables. These fluctuations represent an internal noise--a continuous random disturbance of the orbits of the model from the evolution they would undergo in the absence of the noise (analogous to the well known Brownian motion--the erratically disturbed orbits of pollen grains through the water or air in which they float due to the random impacts of the molecules of the medium). In addition, there will usually be external noise. No useful political system is completely closed--there will always be parts of the real world not included in the system under consideration. The incalculable influence of these external variables is hopefully minor and random, characterized as external noise, and must also be expected to disturb any orbits determined by the model. The net noise--internal and external--acts at each point of an orbit to randomly separate that orbit into a range of possible orbits, much like the observational separation of the starting point.

The necessary existence of noise, which can be represented as small scale random perturbations of the parameters of a model, means that detailed prediction of the system's evolution with such a model is not possible in the chaotic regions of the models' configuration space.

The Richardson model is linear--changes in output are proportional to changes in input--so that initially neighboring orbits remain neighboring; Fig. 1b is valid. But a linear model, such as the Richardson Equation, cannot represent a real world system over all possible ranges of its values. For example, the Richardson model implies that under some circumstances, the values of the military capabilities of the system grow exponentially--they are unbounded with the continued passage of time. Any real capabilities are necessarily bounded--by physical and economic constraints, if not political; they cannot grow indefinitely. Hence, a linear model is at best a limited representation of the real international system, valid for some limited range of values of the variables important in describing the system. Ultimately, policy for a real world system must rely upon non-linear models. Such models predict randomly, and perhaps widely, separating orbits. The existence of such possibilities means that prediction based upon one particular orbit may not have much practical use.

For dynamical models to be useful in dealing with security policies for a real international system, it is thus necessary to develop an understanding of non-linear models with their associated concepts of "chaos", "bifurcation", "complexity". These concepts will be explored in Chapter 4 where it will be shown that it is possible to make useful models of the international system in the face of the omnipresence of non-linear instability. The prediction of chaos in dynamical world models--the prediction of the unpredictable--will itself be profoundly useful.

4. Conclusion

Models, by definition are incomplete representations of a system of interest. In a strongly interacting world in flux, in which the process of change is itself a major factor in the world and its changes, a static model is of doubtful utility--it is unlikely to represent the world well enough. Dynamic models incorporate change and, if non-linear, allow for the impact of the process of change upon the change itself. Even if incomplete in major ways, their creation impels confrontation with the "forces" driving the system. You must go beyond saying the international system prefers minimum interstate "energy", or that peace requires a mutually stable nuclear missile configuration, and address the question: how do we get there from here? If such a dynamic system can be construed to be "close" to reality in some essential way, it can then be used to generate sets of orbits which should be good approximations upon which to base policy. The presence of a transition to chaos may preclude the generation of such useful orbits but at least gives a warning of impending "crisis instability"--itself a very useful aid to policy makers who would attempt to keep their world away from the brink of such crisis instabilities and their strong potentiality for war.

5. References and Additional Readings

Brams, S. 1985. *Superpower Games.* New Have, CT: Yale University Press.

Chadwick, Richard W. 1986. "Richardson Processes and Arms Transfers 1971-1980: A Preliminary Analysis", Journal of Peace Research, 23, No.4: 309-328.

Hill, Walter A. 1992. "Several Sequential Augmentations of Richardson's Arms Race Model". Mathematical and Computer Modelling, 16 (No.8/9): 201-212.

Hill, Walter A. 1992. "Deterministic Quasi-Periodic Behavior of an Arms Race Model". Conflict Management and Peace Science, 12, No. 1. 79-98.

Intriligator, Michael D. and Dagobert L. Brito. 1984. "Can Arms Races Lead to the Outbreak of War?" Journal of Conflict Resolution 28: 63-84.

Kaye, J.D. and G.R. Lindsey. 1970. "MIRVs and the Strategic Balance", Nature **227** : 696-7.

Legault, A. and G. Lindsey 1976. The Dynamics of the Nuclear Balance, Ithaca: Cornell University Press.

Richardson, L.F. 1960a. Arms and Insecurity, Pittsburg: Boxwood Press.

Richardson, L.F. 1960b. Statistics of Deadly Quarrels, Pittsburg: Boxwood Press.

Saaty, Thomas L. 1968. Mathematical Models of Arms Control, New York: Wiley

Schrodt, Philip A. 1981. Preserving Arms Distributions in a Multi-Polar World, Denver: University of Denver Monograph Series in World Affairs, Volume 18, Book 4.

CHAPTER 4
DYNAMICAL CONCEPTS

Since an understanding of dynamical modeling, and its application to international security in a competitive world, is the goal of this book, it is now appropriate to set forth and explain some fundamental dynamical concepts . Some have already been introduced and used; all are included in this chapter for completeness and convenience. The important dynamical concepts are: point, orbit (trajectory), attractor ("strange" and otherwise), basin of attraction, sensitivity to initial conditions and/or system parameters, bifurcation, local vs. system chaos, complexity, linearity and nonlinearity. The political concepts relating dynamical systems to the "real" international systems are: systemic factors, national and system behavior, crisis instability, history, policy and prediction, adaptive behavior. The basic scientific idea of representing a "real system" by a "model" was introduced in Chapter 1, as was the relation between "model" and "theory". The focus of this book is the application of <u>deterministic dynamical law</u> ("theory") to a major problem of international security - the onset of war. A model is created to represent a specific international system or problem based upon assumed specific laws. For a give model, the general parameters of the laws are limited to specific values, and the initial conditions, from which that model is presumed to evolve, are specified. The stability of the model is then examined via the computation of "Liapanov exponents"; a positive value implies instability and the onset of chaos. Using the heuristic conceit that crisis instability in the international system is equivalent to the onset of chaos in the model of that system, it is then possible to evaluate different proposed policies for furthering the stability of the system. All of these notions are set forth in this chapter.

1.Attractors - The Goal of Policy Makers

Policy making implies the choice among different possible starting configurations of the system of interest in order to secure a desirable configuration at some future time. It may also require changing the parameters of the model describing the configuration, which means changing the actual system being modeled, thus making it "run" according to different laws. The system configuration, at any instant of time, is represented by the symbol **x**, which stands for the complete set of numbers (x_1, x_2, x_3,... or x, y, z,...) required to effectively describe the system for the purposes of interest (e.g., x_1 might represent the GNP of one nation, x_2 its military budget, y_1 the GNP of a competitor nation, etc.); the symbol t represents time. To indicate that the system evolves with the passage of time, the system indicator is written as the "variable" $\mathbf{x}(t)$: for every value of t, there is a unique configuration $\mathbf{x}$. The starting, or initial, configuration, at t=0, will be written as $\mathbf{x}_0 = \mathbf{x}(0)$. Thus, in a symbolic graph in which the system time increases along the horizontal axis and its configuration evolves along the vertical axis, the configuration at a specified time, $\mathbf{x}(t)$, is represented by a single point. Different points along a given vertical line on this "configuration space" graph, crossing the horizontal time axis at some specific time value, represent different possible configurations of the system at that same time. As the system evolves, the point describing the system moves through the graph; its path - called an "orbit" or "trajectory" - represents the evolution of the system (Fig. 1a, Chapter 1). Note that any path of realistic interest can only move forward (to the right) in time; traveling back in time is not a realistic option to the policy maker. The role of a deterministic theory is to predict such a path; given any point , $\mathbf{x}(t)$, the theory prescribes the point at the "next instant of time", thus prescribing an orbit over an interval of time. The theory may be sensitive to starting values: different values of $\mathbf{x}_0$ lead to different orbits (Fig. 1a). Different theories (different parameter values in the same mathematical expressions, or different mathematical expressions) may produce different succeeding points for a given point, thus leading to different orbits from a given initial value (Fig. 1b). Finally, different orbits prescribed by a given theory are not allowed to cross each other since <u>the theory is presumed to prescribe a unique successor to each configuration.</u> Thus the paths illustrateed in Fig. 1c are forbidden by deterministic dynamics.

The real world of the policy maker is finite. Hence any numerical variables must remain finite; $\mathbf{x}(t)$ must remain within the finite part of the system-configuration graphs no matter how much time elapses. The configuration variable values may not increase indefinitely; Fig. 2a is prohibited. One allowed possibility is that the orbit approaches a fixed point, $\mathbf{x}_a$ (Fig. 2b) as time increases indefinitely ("goes to infinity"). Such an "attractor" point represents a static configuration: the system evolves to this configuration and remains there, unchanging, in that configuration (e.g., the military exports of each nation in the system remain the same from year to year). Another possibility is that the system eventually cycles back and forth, periodically forever, between two limiting configurations, say $\mathbf{x}_A$ and $\mathbf{x}_B$ (in Fig. 2c). (Many such cycles have been identified in past economic and political behaviors.) In either of these cases, the policy maker in possesion of such a prediction knows exactly what outcomes to anticipate and can act accordingly.

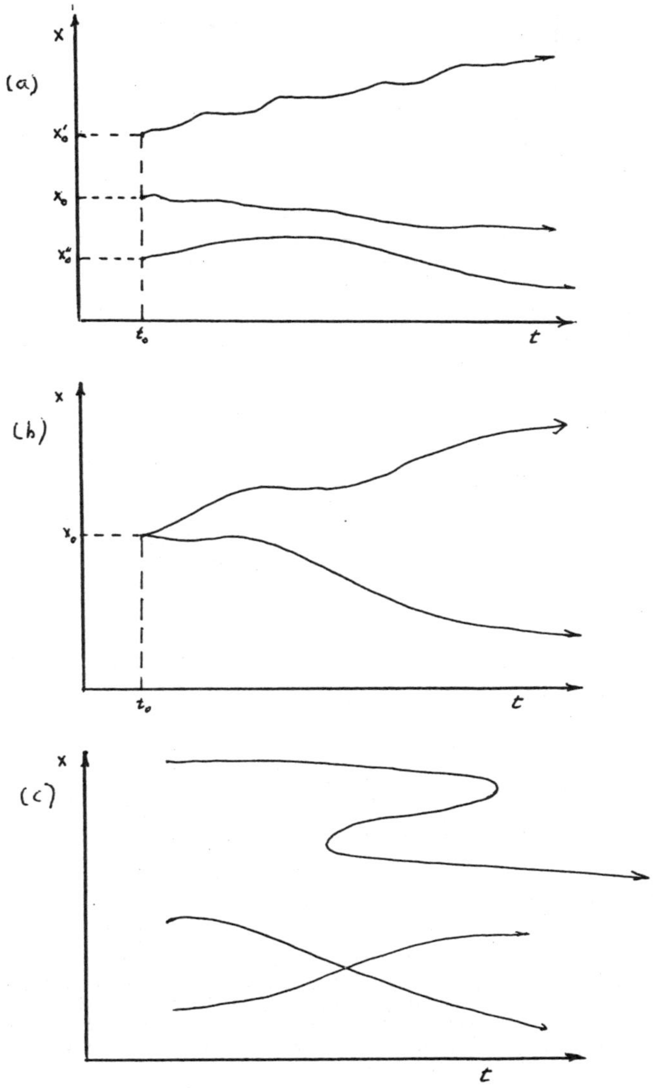

Fig.1. Time evolution of solutions to a dynamical model. (a) Different orbits stem from different initial values under a given set of dynamical laws. (b) Different orbits may start at the same initial point if each evolves under a different set of laws. (c) Orbits may not go backwards in time, nor cross each other, under any set of "reasonable" dynamical laws.

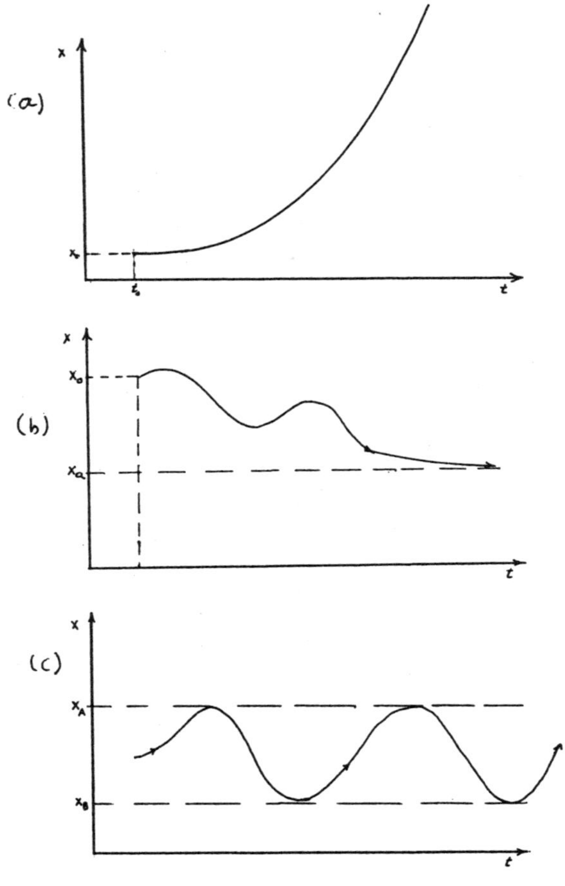

Fig.2. (a) In a finite world, the dependent variables may not increase indefinitely ("go off to infinity") no matter how much time elapses (as the independent variable "time, goes off to infinity"). (b) the dependent variables may approach - get closer and closer to - a fixed point as the time increases indefinitely (the solutions "asymptotically" approach the fixed point). (c) The dependent variables may oscillate regularly between two fixed point ("asymptotic oscillations" or "limit cycle").

As an alternate to the "configuration-space graph" for describing the time evolution of a system, a "phase space graph" may be used. Here, instead of plotting the composite variable **x** against t, the different components of **x**, the different components of **x**, say x_1, x_2 are plotted against each other (e.g., the military prowess of one nation vs that of its opponent). Thus the point attractor in Fig. 2b (in which **x** really represents two variables) has the coordinates x_{A1} and x_{A2} (in Fig. 3a); the orbit gradually approaches the attractor, getting closer and closer to it as time goes on. The "limit cycle" in the configuration-space-graph of Fig. 2b, appears in the phase space graph of Fig. 3b as a closed curve towards which the orbit evolves. In the long-term future the orbit endlessly cycles about the closed curve (e.g., the size of the two competing armies regularly fluctuate up and down indefinitely). Thus far, we have "point attractors" and "line(closed curve) attractors". The job of the dynamical-systems-policy-maker is to determine the attractors of the systems of interest.

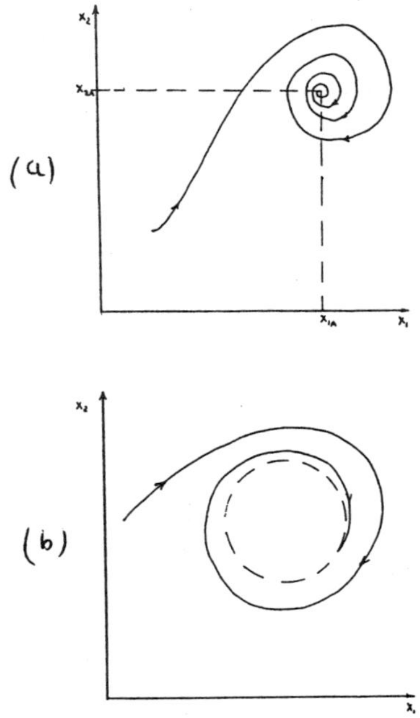

Fig.3. Phase space plots. (a) Oscillating approach to a limit point. (b) Oscillating approach to a limit cycle.

Note that by definition, the attractor represents the configuration (or "cyclic sequence of configurations") which the system reaches "eventually". It is the end point (or curve) which the system reaches as a limiting process and never again leaves. (During finite times the configuration gets closer and closer to the limiting value but may actually not ever "equal" the final attractor value(s). Any point of the configuration-space or phase-space graphs representing a system, is an allowed starting point of the system's evolution, since, by construction, the graphs represent all possible system configurations and the dynamical system laws are supposed to be able to take you from any allowed configuration of the world system to its immediately succeeding (by necessity) configuration. However, any point is not an allowed end point. End points, attractors, are only those configurations for which the succeeding configuration is also an end point. The system laws, acting upon a system represented by a point in an attractor, only can take the system to another attractor configuration. For example, if the system was initially represented by point $\mathbf{x}_A$ in Fig. 3a, it would remain in this configuration "forever"; if it started somewhere in the closed curve attractor of Fig. 3b, it would endlessly cycle about that loop.

Since all points may be starting points whereas not all points may be points on an attractor, there are fewer - usually far fewer - of the latter than of the former. (That is, only comparatively few international configurations can be expected to remain stably fixed or cyclic.) Hence, in a given theory, many orbits, starting from many different initial values, may converge to the same attractor (Fig. 4 a or b). The set of all points which, acting as initial values, evolve to the same attractor, is said to be the "basin of attraction" for that attractor. (Think of all the points on the surrounding hills from which a stone, released to roll downwards without friction or obstruction, would end up at a designated point in the valley floor below; or all the differing possible initial international configurations which will evolve to the same end system.) Since the attractors and their basins represent the evolution of one system, which is governed by the dynamical laws of the system, the attractors and basins are completely determined by (and determine) the laws governing the system.

Because of "continuity" (neighboring configurations are likely to be treated similarly by any "reasonable" set of rules governing the evolution of the system), neighboring points are likely to be in the same basin of attraction, hence they will evolve to the same configuration. There is the possibility, however, that orbits starting as closely neighboring points may end up as radically different end configurations. (Think of two nearby stones starting at the ridge line between two valleys.) In this case - referred to as "bifurcation" - the system is extremely sensitive to slight changes in its initial conditions. A small perturbation in the starting configuration leads to a major difference in the final situation (Fig. 2 of Chapter 1). Such a sensitivity to initial conditions at a given starting configuration would make it impossible for a policy planner to anticipate the

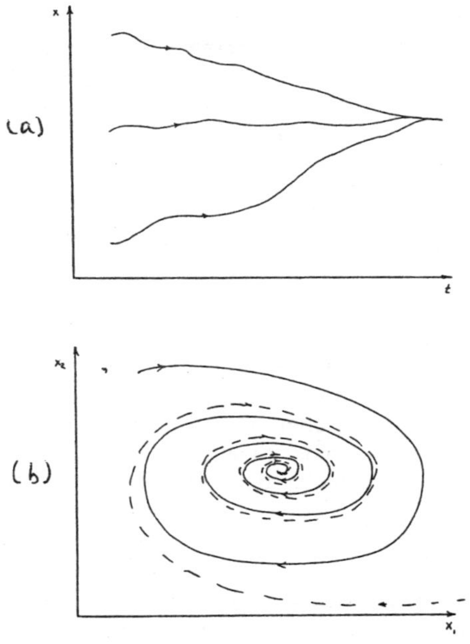

Fig.4. Convergence of different orbits, originating at different initial values, at (a) the same limit point (b) the same limit cycle.

outcome of that system's evolution. However, one could wait a while; the system would evolve away from its initial configuration, on the border between different basins of attraction, and be well into one specific basin. (The stone would be off the ridge line and be definitely in one valley or the other.) The final outcome would then become apparent and policy choices made accordingly. (For example, the Japanese attack on Pearl Harbor may have been such a bifurcation point - on Dec. 7, 1941 it was not certain which side would prevail. However by the Battle of Midway, one-half year later, it was clear that Japan would not win.) Bifurcations delay, but do not seriously hinder the process of prediction and policy making.

The policy maker, of course, does not know which precise point, in the "terrain" corresponding to the system of interest, represents the actual starting configuration of the system. Uncertainties in the measurement process in the real world make such certitude impossible. All that can be determined is a range of values with, perhaps, some associated probabilities. However, if all points in that range belong to the same basin of attraction, the final outcome is still assured.

In addition to uncertainties in starting values, there is also the possibility of uncertainty in the theory itself. The values of the parameters specifying the system law of evolution are themselves the result of empirical observation and consequently are uncertain. However, again by continuity, small changes in system parameters are likely to produce correspondingly small changes in the "terrain" describing the system. Attractors, and their corresponding basins, will change by small amounts, the basic shape of the surface being unchanged. Neighboring initial configurations will still be neighbors, all evolving to the same end point, not that much different from that of the original theory.

Finally, it should be clear that no theoretical model, specified by a finite number of parameters and variables, can give a precise description of a real system. Also no real system of practical interest is completely isolated from the rest of the universe. The effects of the missing descriptions, and of the missing interactions with the remainder of the universe can often be conceptually pictured as random (hopefully small) fluctuations of the model parameters -commonly referred to as "noise". The set of points, representing the connected set of configurations allowed by the laws governing the evolution of the system, can be pictured as a terrain (each surface point representing a possible configuration of the system, each point evolving to a neighboring point), with hills and valleys. We can then describe the noise as a "jiggling of the landscape". Again, it is usually assumed that these small "seismic tremors" do not significantly alter the character of the terrain. Basins, and their attractors, remain essentially the same and so the process of prediction and policy making is unaltered.

1a Mathematical Examples - Continuous Laws of Evolution

Suppose the system is described by a single numerical variable x and it starts with the value x_o . If the system 's dynamical law is such that at subsequent times the variable is

$$x = x_o\, e^{-at} \tag{1}$$

where a is a positive parameter, then - as the time t increases indefinitely - the negative exponential decreases to zero. (Such an evolution would follow for an army whose time rate of decrease, at any time, was specified to be exactly proportional to its size at that time.) Hence the system attractor is $x_A = 0$, its basin of attraction is the set of all values of x. If the time evolution of the system is such that

$$x = x_o \cos \omega t \, e^{-at} \qquad (2)$$

where ω is another positive parameter, the variable still evolves to zero, though now it oscillates through positive and negative values on the way. (The parameter ω determines the rate of oscilation whereas the parameter a fixes the rate of decay.) The point attractor is still $x_A = 0$ and its basin of attraction is again the entire real number line.

Suppose however that the evolution rule is

$$x = x_o e^{-at} + b\,[(e^{x_o t} - e^{-x_o t}) / (e^{x_o t} + e^{-x_o t})] \qquad (3)$$

Now, as t increases indefinitely, x evolves to b if x_o has a positive value and to -b if x_o is negative. There are now two point attractors: +b, whose basin of attraction is all positive values of x_o; -b, whose basin consists of all negative starting values. The boundary point between these two basins of attraction is $x_o = 0$; this is the bifurcation point. No matter how small the magnitude of x_o is (no matter how close it is to $x_o = 0$), if it is positive the system ends up at +b, if negative, at -b. If you are not sure whether the system started with positive or negative values of x_o, just wait a bit. The value of x(t) will eventually be either positive or negative, at which time you will know what the eventual outcome will be.

As an example of a system with three point attractors, three basins, and two bifurcation points, consider the evolution

$$x = x_o e^{-at} + b\,\{[(e^{(x_o-c)t} - e^{-(x_o-c)t}) / (e^{(x_o-c)t} + e^{-(x_o-c)t})] +$$

$$[(e^{(x_o-d)t} - e^{-(x_o-d)t}) / (e^{(x_o-d)t} + e^{-(x_o-d)t})]\} \qquad (4)$$

where c and d are positive numbers such that c<d. If $x_o<c$, the system evolves to $x_A = -2b$; if $c<x_o<d$, the system evolves to $x_A = 0$; if $x_o>d$, the system ends up at $x_A = +2b$. Thus the attractors are: -2b, with basin all x_o such that $x_o<c$; 0, with basin $c < x_o < d$; +2b with basin $x_o > d$. The point $x_o = 0$ is an anomolous point which evolves to a limit cycle, not a limit point.

To illustrate a phase-space evolution, suppose a system is described by two variables, x and y. If the rule of evolution is

$$x = x_A + (x_o - x_A) \cos \omega t \, e^{-at} \quad (5)$$

$$y = y_A + (y_o - y_A) (1 - \sin \omega t) \, e^{-at}$$

the orbit will be as in Fig. 3, the point attractor has the coordinates (x_A , y_A), and its basin of attraction is the entire (x,y) plane. If the law of evolution leads to

$$x = x_o \cos \omega t \, e^{-at} + b \sin \omega t \quad (6)$$

$$y = y_o (1 + \sin \omega t) \, e^{-at} + b (1 - \cos \omega t)$$

the evolution is as in Fig. 3b. The cyclic attractor is a circle, of radius b, whose center is on the y axis at a distance b from the origin. The basin of attraction is again the entire (x,y) plane.

2.Logistics Model - Discrete Law of Evolution

The above examples are "continuous" models; the variables x, y, and t vary continuously. Consider next a discrete model, in which the variables change discretely, e.g., a new budget each year. Variations of this logistic model (commonly used to describe the growth of bacteria, reproducing by binary fission in a resource limited medium) will be used in a subsequent chapter.

Assume the system configuration to be described by a single, positive numerical variable, x, which is bounded above by unity: $0 \le x \le 1$. Assume that x changes from year to year, i.e., x_n is its value in the nth year, x_{n+1} its value in the following year (where n is any positive integer: n=0,1,2....) and the two are functionally related: $x_{n+1} = f(x_n)$ - each configuration is a function of the previous year's configuration. The dynamical law (the "function") governing the evolution of the logistics system is

$$x_{n+1} = r\, x_n(1 - x_n) \tag{7}$$

where the positive system parameter, r, must be less than 4 in order to guarantee that x_{n+1} will be between 0 and 1 if x_n lies between these limits. (The maximum value of x(1 - x) between these limits is 1/4.) Thus, if the starting point, x_o, lies in this unit interval, all subsequent members of the sequence will also be so confined. Given such a starting value, all subsequent values of the discrete variable x_n, can be simply found using a common spreadsheet program on a desk computer. Note that if r > 1 and the second term of Eq.(7), quadratic in x_n, were not present, the sequence would eventually grow beyond all bounds. Thus the quadratic term, x_n^2, is required if x_n is to represent some aspect of a finite world. Any model of the real world cannot be "linear" = its evolution just dependent upon only the first power of x_n.

A point of the sequence will be a "fixed point" if it remains unchanged under the action of the dynamical law: $x_{n+1} = x_n$. Thus the "logistics model" fixed points, x*, are the solutions of the quadratic equation

$$x^* = r\, x^*(1 - x^*). \tag{8a}$$

The only two possible solutions are easily seen to be

$$x^* = 0\ ,\ \ x^* = 1 - 1/a. \tag{8b}$$

From its definition, a point attractor must be a fixed point. But a fixed point need not be an attractor. The additional requirement for an attractor is that all neighboring points must evolve towards it. Thus, if x_n is near to, but not equal to, a point attractor, x_{n+1} must lie closer to that same point. The determination of which, if any of the fixed points is also an attractor can be carried out analytically or graphically. Both methods will be illustrated.

Analytically, let δx_n be the distance between x_n and the fixed point x* while δx_{n+1} is the corresponding distance at the next "year". The functional relation between the two years is

$$x^* + \delta x_{n+1} = f(x^* + \delta x_n) = f(x^*) + \delta x_n\, f'(x^*) \tag{9}$$

where f' is the simple derivative of f with respect to its independent variable (f'(x) =df(x)/dx) and the usual simple derivative approximation has been used in the right side of Eq. (9). From the definition of fixed point, x* =f(x*), it follows from Eq. (9) that $\delta x_{n+1} = \delta x_n f'(x^*)$.

The requirement that the fixed point also be an attractor is that the magnitude of δx_{n+1} be less than the magnitude of δx_n, which implies that $| f'(x^*) | < 1$ is the requirement that x* also be the attractor x_A. For the logistics model, f(x) = r x (1 - x) and f'(x) = r (1- 2x). . Thus the requirement is that $r |1 - 2x^* | < 1$.

Consider first x* = 0 which requires r<1. Thus if 0 < r < 1, this fixed point is an attractor., For this range of r, the other fixed point, x* = 1 - 1/r, is negative, thus not an allowed system point. Hence the point x = 0 is the only attractor for r < 1; its basin is the entire set $0 \le x_o \le 1$. The same result can be obtained (and more easily visualized) graphically. Fig. 5 is the plot of x_{n+1} vs x_n in the unit square. The diagonal line is the locus $x_{n+1} = x_n$; the parabolic curve represents the logistic equation with r <1 (so that its slope at the origin is less than that of the diagonal line). Starting with any value, x_o in the horizontal x_n axis, draw a vertical line to the parabolic curve to get $x_1 = f(x_o)$ projected on the vertical axis.

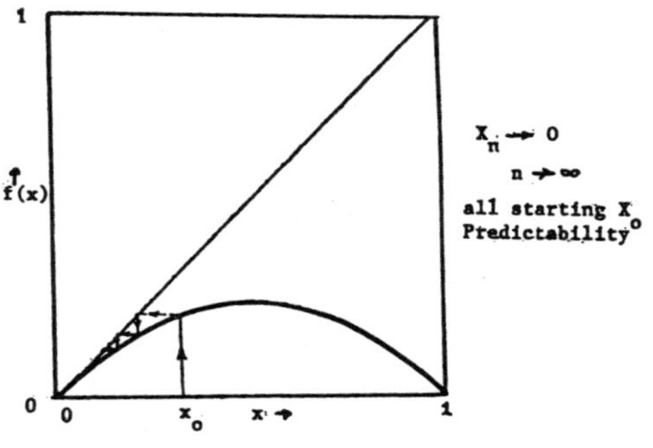

Fig.5. Evolution under the Logistics Model: x_{n+1} versus x_n. The model parameter r is less than one so that the quadratic curve representing the model lies under the straight diaganal line representing $x_{n+1} = x_n$. The single "point attractor" is zero.

At the point where this horizontal projection meets the diagonal line (x_1 now becomes x_n), project down to the curve to get $x_{n+1} = f(x_1) = x_2$. Repeat the process to get x_3, x_4, etc. As is easily seen, as n gets larger and larger, x_n gets closer to zero. This confirms that the point x = 0 is the sole attractor when the logistic parameter r is bounded above by unity.

Now let r > 1, The fixed point $x_o = 0$ no longer satisfies the requirement for an attractor. Substituting the other fixed point into the attractor requirement

$$1 > r \mid 1 - 2x^* \mid = r \mid 1 - 2(1 - 1/r) \mid = \mid 2 - r \mid$$

which is satisfied for 1 < r < 3. Thus the second fixed point is now the sole point attractor for this range of parameter values; again its basin is the entire unit interval., (Note that, as previously said, as the governing laws change, the attractors change.) The slope of the parabolic curve, at the origin, is now greater than one, so the graph is as in Fig. 6. Carefully repeating the previous graphical process, it can be seen that the evolution converges to the point where the diagonal line crosses the parabolic curve, which is the point $x^* = 1 - 1/r$. This is indeed now the sole attractor.

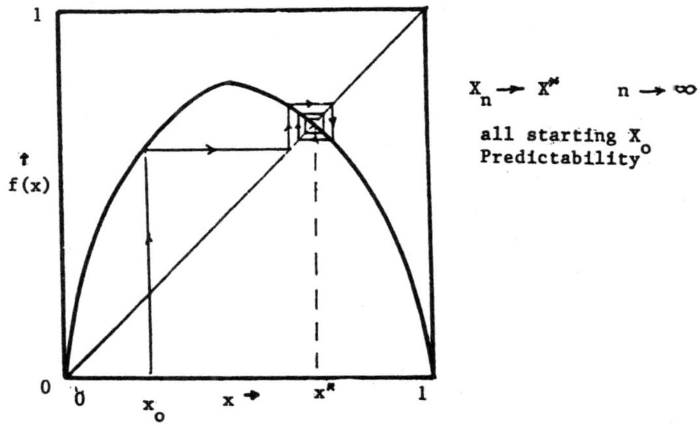

Fig.6. Evolution under the Logistics Model: x_{n+1} versus x_n. The model parameter r lies between 1 and 3 so that the quadratic curve representing the model is cut by the straight diaganal line representing $x_{n+1} = x_n$. The slope of the quadratic curve at the intersection point is less than one so that all allowed evolutions converge to a fixed single point $x^* = 1 - 1/r$, no matter where they start.

When r > 3, the criterion for a point attractor is no longer met. There is no single point towards which the sequence x_n evolves. This can also be seen graphically (Fig. 7). No matter where the sequence starts, it eventually settles down to oscillate back and forth between two values (as in Fig. 2c rather than Fig. 2b which was the case for r < 3). The attractor is now a "limit cycle" rather than a "limit point." Since after settling down to the limit cycle,the evolution repeats itself every two "years" , it is now useful to consider the relation between x_{n+2} and x_n, rather than the previously considered $x_{n+1} = f(x_n)$. Since, in the original sequence, $x_{n+2} = f(x_{n+1})$, the new sequence can be written as $x_{n+2} = f(f(x_n))$, which can now also be examined for fixed points and attractors. The new fixed point equation is a product of two quadratics and hence is a quartic equation having four solutions. By methods similar to those used above, two of these four fixed points satisfy the requirement that they be attractors. They form the limit cycle of Fig. 7: the limit cycle attractor of the original, yearly, evolution is the point attractor of the new bi-yearly sequence.

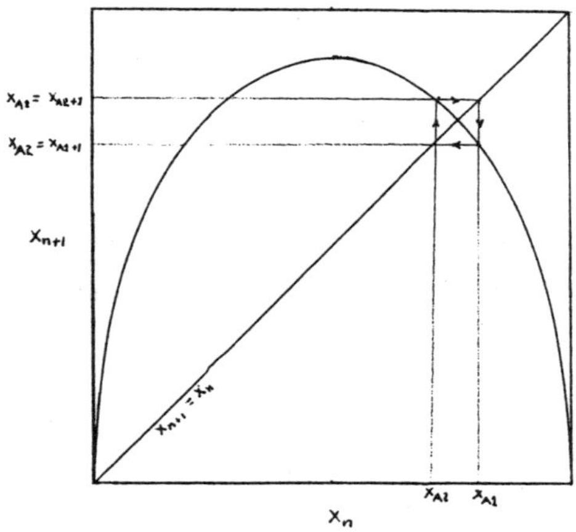

Fig. 7. Evolution under the Logistics Model: x_{n+1} versus x_n. The model parameter r lis greater than 3 so that the quadratic curve representing the model is cut by the straight diaganal line representing $x_{n+1} = x_n$. The slope of the quadratic curve at the intersection point is greater than one so that all allowed evolutions eventually oscillate regularly back and forth between two fixed points x_{A1} and x_{A2} (a "two-point limit cycle").

As the paramter r is further increased, to somewhere between 3.4 and 3.5, the fixed points of the bi-yearly sequence cease to satisfy the point attractor criteria. It no longer has a point attractor which means that the original yearly sequence no longer cycles back and forth betwen two points. It now cycles between four points (as in Fig. 8)! A similar argument shows that this four-point-cyclic attractor is a manifistation of the one-point attractors in the four-year sequence $x_n, \ldots, x_{n+4}$: $x_{n+4} = f(f(f(f(x_n))))$. This four point cyclic attractor breaks down, and is replaced by an eight-point cycle attractor when r is increased to somewhat beyond 3. Further increases in the parameter r lead to period-16 cycles, period-32 cycles, etc. Although cycles (groups of points) rather than single points, these attractors still share the fundamental characteristic of attractors: all starting points in their basin of attraction - in this case the entire unit line allowed to the system - evolve towards the same cycle, around which they will circle endlessly. Since the cycle is of finite length, no two points on it can be very far apart. This means that all orbits, no matter how far apart their initial points, will essentially be traveling together at "the end". prediction is still possible - the final configuration will cycle endlessly among a number of well-defined, similar possibilities.

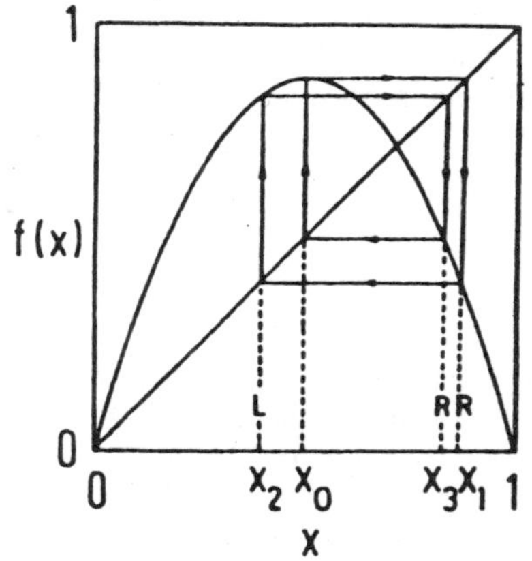

Fig. 8. The logistics Model with a four-point limit cycle.

By the time the system parameter reaches the critical value $r_\infty = 3.569944\ldots$, an infinite number of period doublings have occurred. A fundamental change occurs in the systems behavior at this critical value of r; beyond it lies "chaos".

3.Chaos

We have seen that the ability to predict the evolution of a given starting configuration depends upon the proximity of the point representing that configuration to the basin boundaries determined by the deterministic laws governing the system. Suppose that the basin boundaries are such that every point in the region of interest is arbitrarily close to a boundary. Given such an, apparently bizarre, law of system evolution, orbits starting on neighboring points may diverge from each other no matter where these points lie. All points are bifurcation points so that all groups of points, no matter how closely neighborly they are, evolve to different atteractors. Prediction of eventual outcome would then be impossible. Such a situation, where every point is extremely sensitive to initial conditions, is referred to as "chaos", the deterministic evolution is "chaotic". The set of attractor points for such an evolution is called a "strange attractor". Any two starting points in such a system give rise to a pair of orbits which may end up widely separated on the strange attractor. A good physical example of such chaos is a turbulent brook; the ever present "noise" means that the water flow at any time and place is constantly changing in a completely unpredictable manner. The stock market, in a time of great uncertainty, may be another example of chaos, as is the uncertain melee on the battlefield referred to as "the fog of battle" by von Clausiwitz.

3a.Example: (The Cantor Set)

Suppose a unit length straight line is divided into three equal parts (Fig.9). Assume that the dynamical law is such that the division points fall in one basin of attraction; the remaining non-division points constitute a second basin. Now divide each of the three segments into three further subsegments. Again, all division points lie in the first basin, non-division points in the other basin. Repeat the process over and over again, dividing each segment into three sub-segments. As the process continues indefinitely, the division points get arbitrarily close to each other, as do the non-division points. Any arbitrary collection of points on the line, no matter how compact, will contain points in each of the two basins of attraction, hence will give rise to diverging orbits.

Fig. 9. Cantor Set. Straight line of unit length devided into three equal parts. Each part is then devided into three equal parts, and the process is repeated indefinitely.

3b.Liaponov Exponents.

In the above example, the set of bifurcation points, implying chaos, is given. Normally, such information is not immediately obvious. Instead you are given the law by which the system of interest evolves, e.g., the logistic law of Eq. 7. From this you must determine which, if any, regions of possible starting points, and which ranges of systems parameters, are chaotic. This information is necessary for the policy maker to know what kind of predictions can be confidently made.

The defining characteristic of chaos in a dynamical system is "sensitive dependence upon initial conditions" everywhere in a range of initial conditions. (Chaos must be distinguished from a simple bifurcation, where such sensitivity occurs at a single, isolated point.) The most direct method for seeing whether or not chaos is present is to examine large numbers of orbits, starting as close neighbors, and see whether or not they diverge from each other. Such a universal divergence is the measure of chaos.

Consider, for example, the logistic sequence, Eq. (7). Take an arbitrary starting point, x_o, in the range of interest, another nearby starting point, $x'_o = x_o + \delta x_o$ (where δx_o is a small random nunmber), and follow the two sequences evolving from these two initial values: x_n, x'_n. (A simple spread-sheet computation can produce both sequences simultaneously.) The separation between the two orbits is $\delta x_n = x_n - x'_n$ and so the requirement for chaos is that the ratio $| \delta x_n / \delta x_o |$ grows beyond all bounds, as n increases indefinitely, for all x_o in the region of interest and for a large collection of δx_o values.

Growth is most easily represented as an exponential, i.e., we expect that, for large enough values of n,

$$\delta x_n \sim \delta x_o \, e^{\lambda n} \tag{10}$$

where λ is a parameter characteristic of the evolution law - the "Liapunov exponent". Positive λ implies growing separation and chaos. From its definition, Eq. (10), the exponent is obtained as

$$\lambda = (1/n) \ln | \delta x_n / \delta x_o | \tag{11}$$

which is easily calculated using the spread sheet routine. Since we are interested in whether or not there is an "attractor", the calculation in Eq. (11), must be repeated for a number of large values of n. The Liapunov exponent is that value of λ which results when λ no longer changes as n is further increased. ("λ is the limit of the expression on the right of Eq. (11) as n goes to infinity"). If the concept is valid, λ should not change as δx_0 is randomly varied (though kept small). From the defintions given (both theoretical and practical), it should be clear that negative exponents mean that initially neighboring orbits converge to attractors whereas positive values of λ imply exponential divergence of such orbits - hence chaos.

Carrying out this process for the logistic law of system evolution, Eq.(7), for a series of values of r, a plot of Liapanov exponent versus systems parameter is obtained (Fig. 10b). Above it (Fig 10a), for the same range of r, is plotted the location of the effective attractor points - those points to which the sequence keeps returning, over and over again, as the system "time" n increases endlessly. In this latter figure it is easy to see the single point attractor region (in which all orbits end at one point) for $r < 3$, the two-point cycle attractor region, the four-point region, etc., up to the critical value r_∞. The Liapunov exponent is not positive in any of these regions, confirming that these points - single, or cyclic - truly are attractors. But in the remainder of the allowed parameter region, $r > r_\infty$, the exponent is primarily positive - most of this region is chaotic. (Note that there are "islands" of non-chaos imbedded in the chaotic sea). This chaos interpretation is confirmed by noting, in Fig. 10a, that the orbits never end, nor do they cycle back and forth, in a definite order, among a fixed number of points - they wander all over a region in a never-ending manner. Ordinary prediction is impossible in that an arbitrary orbit never ends and a pair of such orbits, initially neighbors, always diverges exponentially.

Not only is there a "sensitive" dependence upon initial conditions for those values of the parameter r for which chaos is indicated, there is also a "sensitive dependence" upon the value of the system parameter : one value of r may produce a chaotic evolution of the sequence whereas a very close neighbor value may lead to a very different non-chaotic sequence. Thus small fluctuations in either starting values or system parameters - both expected in any attempt to model a real world system - may lead to very large, unanticipated, changes in system configuration. If the behavior of the simple logistic equation, which is relatively easy to determine, is anything like that of the more complex evolution rules, introduced in subsequent chapters, that may determine significant aspects of the real world system, prediction of specific outcomes may be very difficult indeed.

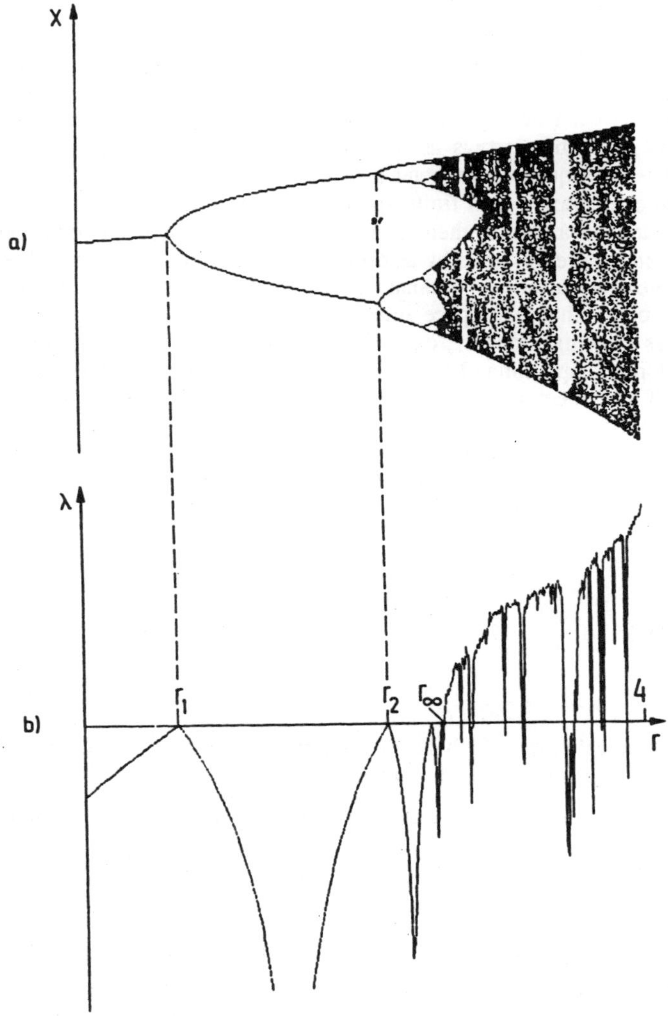

Fig.10. (a) The attractor point of the Logistics Model plotted against the increasing value of the model parameter r. The areas dense with point represent chaos. (b) The values of the Liapunov exponents for the Logistics Model versus the model parameter. Positive exponents signify chaos. (From Schuster,1988; used with permission of the publisher.)

3c. Strange Attractors.

It is clear, in the chaotic regions of Fig. 10a, that any single orbit ranges over a finite region of variation of the system variable x, which seems to imply that two such orbits can't be separated by more than a finite distance., Yet, in this region, their separation grows as a a positive exponential, Eq. (10), which is unbounded as the sequence parameter n grows beyond all bounds. How can there be an infinite separation in a finite space? If we say that each of these orbits is tending towards an attractor, then it must be a very strange attractor indeed. It certainly cannot be represented by an ordinary single point, collection of points, or simple curves on surfaces. Ordinary "dimensionality" is no longer valid - the "curve" is longer than the space into which it must fit. It perhaps can be visualized as a "wiggly curve", one with an infinite number of wiggles, each of zero size. On perhaps it may be seen as occurring in many parallel, superimposed levels, as if the sheet upon which it is drawn actually consists of an infinite number of sheets, lying one-on-top-of-the-other. It is often referred to as a "fractal" - it has a non-integer "dimension". (Readers intrigued by such notions are advised to pursue the matter in some of the more mathematical works cited in the references at the end of this chapter.)

4. Linearity

The dynamical rules which govern our system create a unique orbit, for each allowed starting point $\mathbf{x}_o$, which connects that starting point to its corresponding configuration $\mathbf{x}(t)$ at the future time t. This orbit is symbolically written as

$$\mathbf{x}(t) = f\,(\,t, \mathbf{x}_o, \mathbf{a}\,) \tag{12}$$

where $\mathbf{a}$ represents the set of parameters which specify the governing rules. We say that $\mathbf{x}$ is a "function, f, of t." (The examples given in Section 1a of this chapter illustrate the function concept; e.g., for Eq. (1), $f = x_o\, e^{-at}$.) This function is <u>the solution</u> to the equations representing the governing rules, for the initial vlaue x_o. For a fixed set of rules (fixed $\mathbf{a}$), a different starting value, $\mathbf{x}_o'$ implies a <u>different solution</u> to the <u>same dynamical equations</u>:

$$\mathbf{x}'(t) = f\,(\,t, \mathbf{x}'_o, \mathbf{a}\,) \tag{13}$$

Consider now a new function of time, y (t) which is the sum of the two given solution functions, (12) and (13):

$$\mathbf{y}(t) = \mathbf{x}\ (t) + \mathbf{x}'\ (t). \tag{14}$$

Suppose a system point $\mathbf{y}_o$ can be found such that the orbit starting from it, determined by the assumed dynamical equations, is equal to $\mathbf{y}$ (t), that is,

$$\mathbf{y}\ (t) = f\ (\ t, \mathbf{y}_o, \mathbf{a}\ ) \tag{15}$$

i.e., $\mathbf{y}$ (t) is another solution to the same dynamical equations. Then these dynamical equations are said to be "linear". If $\mathbf{x}_1$ (t) and $\mathbf{x}_2$(t) are any pair of solutions to a set of "linear equations", it follows from the definition that the function

$$\mathbf{z}\ (t) = a\mathbf{x}_1\ (t) + b\mathbf{x}_2(t) \tag{16}$$

where a and b are arbitrary constants, is also a solution to that same set. ($\mathbf{z}$ (t) is said to be "a linear combination of $\mathbf{x}_1$ (t) and $\mathbf{x}_2$(t).") Using the "distributive law of multiplication", $a\ (x \pm y\ ) = ax \pm ay$, and the linearity of the ordinary calculus derivative

$$(d/dt)\ (\ ax + by) = a\ (dx/dt) + b\ (dy/dt),$$

it follows, for example, that the Richardson Equations of the previous chapter are linear: the Richardson Model is a Linear Model. On the other hand, because of the $x_n^{\ 2}$ term appearing in Eq. (7), the Logistics model is not linear: if

$$x_{n+1} = a\ x_n(\ 1 - x_n\ ) \qquad \text{and} \qquad x'_{n+1} = a\ x'_n(\ 1 - x'_n\ )$$

then it does not follow that

$$x_{n+1} + x'_{n+1} = a\ (x_n + x'_n)(\ 1 -\ x_n - x'_n\ ).$$

As will be shown shortly, linear equations do not have bifurcation points. Thus they can have at most one attractor, one basin of attraction. (If the Richardson model is "stable", it has one point attractor; otherwise, it has none.) Having no bifurcations, chaos is not possible for a linear model. Orbits which start as neighbors remain as neighbors. Prediction is always possible if the system is governed by linear deterministic laws.

4a.Is the Whole Just the Sum of Its Parts?

Consider a system made up of a number of identical parts, each subject to the same rules governing its evolution. If these rules are linear, then a knowledge of how each element evolves, without simultaneously considering the behavior of all the other elements, is sufficient to determine how the sum of all of these elements evolves. In other words, the behavior of each element, following its prescribed rules, is independent of whether or not the other elements are present. In as much as its behavior defines the existence of an element, elements of a linear system exist independently of each other's existence.

For example, in the Richardson Model, let s(t) be the sum of the collection $x_1(t), x_2(t)...x_n(t)$, whereas y (t) is the sum of $y_1(t), y_2(t)...y_n(t)$; $x_1(t)$ and $y_1(t)$ are related via the Richardson Equations (Ch.3, Eq. (1)), as are $x_2(t)$ and $y_2(t)$, etc, all with the same set of parameters. Then since the Richardson Equations are linear, x(t) and y(t) also satisfy the same set of equations. If you know how each pair, x_i (t) and y_i (t), is evolving, independently of all the other elements, then you know how the collective evolves. Thus the whole is simply the sum of its parts.

Conversely, if the sum of solutions, each representing an element evolving under the system's dynamical equations, is not itself a solution (if the system is "non-linear"), then the behavior of the system is not simply determined by the behavior of its constituent parts.

In a linear system, the individuals obey the same evolution rules as does the system; if one is known, so is the other, irrespective of the behavior of other elements of the system. Thus the individual can make "rational choices" on the basis of the system's rules without having to take cognizance of the behavior of other elements of the system. Thus the "rational choice" models of the political scientists are analagous to linear system models.

On the other hand, in a non-linear system, knowing the rules by which the sum of all system elements evolve is not sufficient to determine the rules governing the evolution of a specified element. Knowledge must be obtained about all other evolutions to obtain the desired evolution by subtraction. Thus the behavior of a given element is "adapted" to the presence and behaviors of other elements. It follows that "adaptive behavior" is the appropriate political science metaphor to describe non-linear systems.

4b. Proof by Contradiction: Linear Systems Do Not Bifurcate!

Consider a one-dimensional linear theory, i.e., there is a single numerical dependent variable x(t). Suppose the theory has a bifurcation point; for simplicity, let it be at x = 0. Thus all negative starting points, $x_o < 0$, evolve towards the point attractor x_A; all positive starting points, $x_o > 0$, go to the different attractor x_B. Let $x_1(t)$ and $x_2(t)$ be two solutions to the linear equations governing the system starting on either side of the bifurcation point: $x_1(0) = -\varepsilon$, $x_2(0) = +\varepsilon$, where ε is an arbitrarily small, postive number. Then $x_1(t) \rightarrow x_A$, $x_2(t) \rightarrow x_B$. Since the theory is linear, an arbitrary linear combination of x_1 and x_2, $x_3(t) = ax_1(t) + bx_2(t)$, is also a solution to the system equations. Letting t be the starting time, t = 0, $x_3(0) = ax_1(0) + bx_2(0) = (b - a)\varepsilon$. If b> a, x_3 starts in the positive basin: $x_3(t) \rightarrow x_B$. But, from its definition, $x_3(t) \rightarrow ax_A + bx_B \neq x_B$ if $a \neq 0$. Thus, $x_3(t)$ cannot be a solution, i.e., the equations cannot be linear if there is a bifurcation.

5. Non-Linearity - Must Be Present in the Real World

We have seen that the linear Richardson model sometimes has a solution which requires an indefinite growth of its dependent variable (positive exponential function of an indefinitely growing independent variable - the time). It eventually runs counter to "reality" in a finite world. Thus linearity must eventually break down. Something like the quadratic term of the logistics model must become a signficant factor in any theoretical model which is expected to be a useful adjunct to realistic policy making. But once such a "non-linearity" comes into play, the usual ability to make predictions may disappear - chaos may be present.

Given a specified non-linear theory, with its inherent possibility of producing bifurcations, Liapanov exponents can be calculated for "typical" points in those ranges of system parameters and initial values which may be of practical or theoretical interest. If the resultant exponents are negative, the theory is "stable" in those regions - it is not unduly sensitive to conditions - and ordinary predictions can be made to help guide policy making. If the calculated exponents are positive, the system is sensitive to conditions and ordinary predictions can not be made. But the theory is not useless to the policy maker. The theory has predicted chaos - itself a potentially useful adjunct to policy making, as will be seen subsequently.

Political scientists have introduced the concept of "crisis instability" (e.g., Brecher and Wilkenfeld, 1997) to describe a political configuration which is very sensitive to perturbations. For example, the relationship between the U.S. and Spain at the close of the 19th Century, and the internal political situation in the U.S., was such an unstable configuration. The response to a relatively minor perturbation, the destruction of a single U.S. Ship, the U.S.S. Maine, in the Spanish-controlled port of Havana, Cuba, due to circumstances which may well have been accidental, or not subject to Spanish Governmental control, led to a disproportionately large response - the Spanish-American War with its significantly major changes in the world status of Spain and the U.S. Conversely, there are "crisis -stable" configurations: significant insults to a relationship do not lead to the disruption or even signifciant change, in that relationship. Cold War examples abound in which, e.g., aircraft belonging to NATO or "western" affiliated nations were fired upon, or destroyed with significant loss of life, by WTO or "eastern" nations. Verbal vituperation often followed, but the violence remained quite constrained. <u>The analogy between the political-science concepts of stability and instability and the dynamic modeling concepts of insensitivity or sensitivity to initial conditions should be quite evident.</u> Assume it important to the policy maker to know whether the world he confronts is, or will be, stable or unstable in response to his anticipated changes. It should therefore be very useful to be able to know, of a dynamical model which is a "fair representation" of that world, if, and where, chaos is, or is not, likely.

Examining Fig. (10b), it is evident that the extent of the region over which "chaotic motion" can occur, the "size" of the strange attractor, may vary as the system parameter r is varied. For some system configurations and/or laws, the extent of the strange attractor is small compared to the allowed range of the system variables. This will be referring to as "local chaos" or as "weak chaos". The system is known to end up in a small part of the allowed system range, though, of course, where in that region it will "end up" cannot be predicted. At the other extreme, which will be referred to as "system-wide chaos" or "strong chaos", separations on the strange attractor may have the same extent as the entire range of possibilities allowed to the system. You know nothing at all about the future evolution of such a system.

"Local chaos" means relatively small scale random fluctuations. (Physically, this would be exemplified by micro-turbulence, small-scale random ripples upon a basically well-ordered body of water.) It would still be possible to average over such small random fluctuations to obtain an overall prediction. For example, it may not be possible to fortell which individual soldier will be killed in a specific battle, but still be possible to predict the battle's outcome; or, it may not be possible to predict who will win a given battle but still be able to pick the winner of the war. Basic "foreknowledge" of the system behavior is possible, providing the relevant policy makers with the ability to exert basic control.

The system wide fluctuations of strong chaos means that the model provides no foreknowledge at all as to what will happen. The smallest disturbance may produce an undamped explosion of the entire system, e.g., the 1914 revolver shots at Sarejevo which blew up the European world.

6. Complexity and History

A system whose evolution is well represented by a linear model (such as that of Richardson), or by a non-linear model in its stable region, is "explained" by that model. That is, the system configuration at any time is uniquely predicted (within "experimental errors") via the given ("measured") system values at some (perhaps long-distant-past) initial time. There is no need to become aware of the system's configurations at intermediate times. The systems future can be predicted, i.e., rational policy can be made, just on the basis of knowlege of its past. Even though the real system may have displayed fluctuations about the predicted orbit all along, its basic evolutionary trends were set back in the past. They may be relied upon to continue due to the basic insensitivity of the system to small disturbances. The policy maker faces no unanticipated loss once the path has been set (the policy has been chosen) and so need not make extensive provisions for "reserves" to "cover" such losses.

The situation is quite the opposite if it requires an unstable non-linear model for its representation. Knowledge of yesterday's conditions may be of little use in predicting tomorrow's events in view of the possibility that small, but significant, disturbances are occuring today, even as the predictions are being made. Even exponential growth is not completely "explosive": information about today's system may let you "see" a short time into the future. But long distant prediction and policy making can not be rational. (The dinner party example of Chapter 1 is a good metaphor for this situation. For a physical analogy, consider a pin-ball machine: the route of the ball is completely predictable between its collisions with the bumpers, but the outcome of each successive collision with an obstacle on the machines playing surface is completely random. Hence the final destination of the ball, in an undisturbed machine, is completely unknown at the start. As a similar example from macro-economics, the extensive buying of its own currency by a national bank may stabilize the international exchange value of that currency for the near future. It may prevent, or lead to, complete collapse of the currency in the long run.) Any "explanation" of today's events in such a system requires complete knowledge of its immediate as well as past history. If the system instability is local, the policy maker must have sufficient reserve resources - and flexibility of tactics - to be able to make up for unanticipated losses (e.g., money,men, or battles). The creation of such reserve is a change in the system and the model representing it, leading to a shifting of the attractors, their types, and their basins.

If the chaos is system wide, complete flexibility of strategy is required as well as effectively infinite resources in reserve. Even given these, there can be no rational expectation of control of the outcome. Policy makers can either try to change the system (alter its parameters) so as to make it either locally chaotic and favorable, when averaged, or predictably favorable; or they make take actions with unpredictable consequence. In the latter case, they can only hope that their chosen activity will be no worse than its alternates or to no activity. (For example, the Japanese Government, as it embarked upon its Pearl Harbor adventure, may have had no hope of actually defeating the U.S. in a prolonged war. The presumed rational hope was for a stalemate or a defeat which would leave them no worse off than the continuation of their existing "pre-war" situation. Even so, the Japanese influenced a change in the system by getting their German allies to declare war on the U.S. Thereby, the system was changed from a two-component system, in which the U.S. would have used its entire resources against Japan, to a multiple-actor system, in which the U.S. initially diverted a major share of its rapidly growing military resources to the defeat of Germany.)

Hence, even though the prediction of specific outcomes may not be possible in a non-linear model - such as is expected to be required in order to represent a real international system reasonably well - the ability to predict chaos and its extent, the ability to predict unpredictability and its character, may still be a very useful tool in policy making. Examples of the use of such "unpredictability predictions' will be given in the following chapters.

7. References and Additional Readings

Axelrod, Robert M.1997. *The complexity of Cooperation: Agent Based Models of Competition and Cooperation.* Princeton, N.J.: Princeton University Press

Beyerchen, Alan 1992. "Clausewitz, Nonlinearity, and the Unpredictability of War", *International Security* **17:3,** 59.

Brecher, Michael and Jonathan Wilkenfeld. 1997 *A Study of Crisis.* Ann Arbor, MI: University of Michigan Press

Brecher, Michael 1989. *Crisis, Conflict, and Instability.* New York: Pergamon.

Casti, J.L. 1994.*Complexification: Explaining a Paradoxical World Through the Science of Surprise.* New York: HarperCollins.

Casti, J.L. 1992. *Reality Rules.* New York: Wiley.

Casti, J.L. 1990. *Searching for Certainty.* New York: William Morrow.

Ekeland, Ivar 1993. *The Broken Dice.* Chicago: University of Chicago Press.

Gleick, James 1988. *Chaos: The Making of a New Science.* New York: Penguin.

Mann, Steven R. 1992. "Chaos Theory and Strategic Thought", *Parameters,* Autumn, p54.

Nicolis, G. 1989. *Exploring Complexity: An Introductory Text.* New York: W.H. Freeman.

Ott, E. 1993. *Chaos in Dynamical Systems.* Cambridge: Cambridge University Press.

Pearson, Frederick S. and J. Martin Rochester 1998. *International Relations,* 4th Ed. New York: McGraw-Hill.

Prigogine, I and I. Stengers 1984. *Order Out of Chaos. New York" Bantam.*

Prigogine, I and I. Stengers 1997. *The End of Certainty: Time, Chaos,and the New Laws of Nature.* New York: Free Press.

Saperstein, Alvin M. 1997. "Complexity, Chaos, and National Security: Metaphors or Tools?" in *Complexity, Global Politics and National Security,* David S. Alberts and Thomas J Czersinski, Eds. Washington, D.C.: National Defense University.

Saaty, Thomas L. 1968. Mathematical Models of Arms Control, New York: Wiley

Schuster, Heinz George. 1988. *Deterministic Chaos - An Introduction:* Second Revised Edition. Weinman, FRG: VCH Verlagsgesellschaft mbH.

Waldrop, M.Mitchell 1992. *Complexity: The Emerging Science at the Edge of Order and Chaos.* New York: Simon and Schuster.

CHAPTER 5

CHAOS AND INTERNATIONAL STABILITY — MODELING WAR INITIATION

Having introduced the notions of "dynamical modeling", it is now appropriate to attempt to apply them to the understanding of important problems facing the international system. Given the apparent fragility of our world in the face of "all out nuclear war", and the ever-present possibility that "ordinary" international war might lead to such a nuclear disaster, trying to understand the conditions which may lead to, or prevent, the outbreak of war would seem to be an excellent application of the new ideas. Certainly, there have been many previous attempts to apply some aspects of dynamical modeling to the problem of the initiation of war (e.g., references in previous chapters such as Casti 1990, Hill 1992, Richardson 1960a.b. Saaty 1968; Schrodt 1981). The practical goal is the application of the new knowledge - an extention of the policy-making technology - to the avoidance of international catastophe. Thus, any indicator of impending war (or crisis instability) would be a useful sign of undesirable policy.

It may be, as maintained by Kennedy (1987), that relative economic strength is a good indicator of the outcome of a race for dominance in a competitive world, but this does not imply that war itself is a necessary stage in achieving dominance of a given system (or even that victory in war leads to dominance on the international stage). It is the outbreak of war with which I am concerned here. It also may be that, given war, relative "economic might" determines the war's outcome, at least probabilistically. But probability is not certainty (otherwise there would be no need for the war and its consequent losses, even to the winner); there are many historical examples where the "winner" is not the dominant economic power. A dynamical model, to be useful, should predict the outbreak of war (whether or not there is a chance that it could predict the outcome of the war).

1.Prediction and Policy Making

To repeat what has been said in previous chapters, creative technology—the control of significant aspects of one's world—implies the ability to predict the outcome of a design, construction or deployment strategy. In the physical sciences, such prediction requires a mathematical model based upon theoretical understanding. The political and policy "sciences" obviously covet such a level of theoretical understanding, an understanding which would allow the prediction of outcome for each policy alternative and hence the ability to choose policy to match with desired outcome. A mathematical model allows all postulates and logical steps of such understanding to be explicit. Thus, let X_0 represent the present state of the international system under consideration. Policy choices will change that state into a different state X_n (at future "time" n); different policies result in different futures, i.e., different values of X_n. Prediction implies knowing X_n, given X_0 and a specific policy. This implies a unique relationship, for a given policy, between X_0 and X_n, i.e., each "stage of history" determines the next stage.

This determination is expressed mathematically as a function:

$$X_{n+1} = f(X_n, \lambda) \tag{1}$$

where λ represents the collection of parameters which define the form of the function *f.* Different policies imply different functional relations between present and future so that different choices of λ correspond to different policy options or national characteristics. Eq. (1) relates the state of the system at "time" n to the next time n + 1: the next year, the next budget cycle, or whatever unit is most convenient for describing the evolution of policy and events. To go a long way into the future, say from 0 ("now") to n, Eq. (1) is iterated n times: $X_1 = f(X_0,\lambda)$, $X_2 = f(X_1,\lambda)$, $X_3 = f(X_2,\lambda)$, etc. The final outcome of any policy is the value of X a very long time in the future: $X_\infty = X_n$ as n becomes infinite. This value, or set of values, is called the "attractor" of Eq. (1), and varies as the parameter set λ is varied. Thus the problem of prediction becomes: given X_0 and λ, what is the attractor X_∞? The job of the policy maker is to either change the present state of affairs, X_0, and/or the policies, λ, connecting the present to the future, so as to ensure a desired end state X_∞, or to avoid an undesirable attractor (Saperstein 1990).

It is certainly true that many random events affect the future development of the world system made up of many nations, each in turn comprised of many millions of complex human beings, all subject to an occasionally capricious nature (hurricanes, earthquakes, etc.). Such randomness cannot be represented by determinate relations such as Eq. (1). If randomness were dominant, if the variations in possible outcomes due to it were greater than the variations predicted by Eq. (1), no prediction would be possible and world policy makers would be reduced to reading goat's entrails. <u>Most modern policy makers proceed under the working assumption that they have some control over the future—that random</u>

events are a minor add-on to the "if- then" deterministic equations above—even if such equations ate only implicit in their minds. It seems reasonable for international relations policy makers and scholars to assume that in spite of the apparently random behavior of individual members of their populations, nations interact with each other following patterns or "laws" which can be surmised and used in policy making; e.g., "The initial steps on the road to war were taken with a feeling of *necessity*,..." (White 1985, p.234; emphasis added). *Necessity implies law.*

There are those, of course, who argue against the usefulness of modeling the world system as a system under law, Eq.(1), for the understanding of the outbreak of war in the system. It is the specific situation, not the system, which must be understood; thus a "situation analysis" is called for, one which emphasises the role of the individual leader: "The importance of personality characteristics in a crisis is clearly an important and complex question..." (Farrar 1985, p.247). I certainly believe that there are situations where a dominant leader, e.g., Hitler, welcomes or even seeks war. But that personality may not be able to bring about, or avoid, war independently of the structure of the given world system; it may be "...that roles are a more important consideration .... that the decision-makers were in a sense 'programmed' for the crisis and war ..." (Farrar 1985, p.247).

The dichotomy of view is clear: "Whereas situation analysis assumes that crisis behavior is due largely to crisis conditions, system analysis assumes that crisis behavior is largely attributable to previous conditions. For the first crises are the cause; for the second they are the effect" (Farrar 1985, p.248). In this book, a system view is adopted: "...conflicts will recur as long as the existing system persists" (Farrar 1985, p.248). The system analysis is carried out by modeling the system via explicit dynamical mathematical models.

2. Predicting War

The initiation of a major war represents a fundamental change of state of the international system, as distinctive as the change of a water system from a solid state (ice) to a liquid state. A model capable of predicting such a war should thus contain within it the possibility of more than one kind of system state and the possibility of transitions between such states. Such a change of state is not apparent in any of the common static or dynamic arms race or international security models such as those illustrated in Chapters 2 and 3. (1 am not concerned here with the large game theory literature on these questions, e.g., Brains, 1985.) They give no indication of "crisis instability."

Crisis instability (cf., e.g., Brecher and Wilkenfeld, 1997) means an extreme sensitivity of a political system's behavior to small changes in input or system parameters. A good example of such an instability is Europe early in this century: the shooting of an archduke in the Balkans was a minor insult compared to the mass slaughter of World War I which followed (e.g., cf Tuchman, 1962). Conversely, war did not follow the much greater insults contained in the Cuban Missile Crisis (E.g., Allison,1971), nor did the shooting down of a Korean 747 jetliner (KAL007) by Soviet jets, with the consequent loss of hundreds of lives, escalate to further murders (e.g., Reagan 1983). Thus the Eurasian system was significantly crisis unstable at the beginning of this century but comparatively stable in its latter half when small system perturbations led to commensurately small changes in the system output. It thus appears that some configurations of an international system may be stable, others crisis-unstable. In the language of Eq. (1), there are some ranges of λ values for which small changes in X_0 lead to correspondingly small changes in X_∞, other ranges for which small deviations in X_0 may produce wildly disproportionate shifts in X_∞ -- in short time intervals X_n takes on all possible values.

Assuming that minor perturbations - such as insults, assasinations, attacks on embassies, aircraft, or ships - will always be present in a system of competing states and nationalisms, the presence of *crisis instability - the inability to predict and hence control, the outcomes of such insults and prevent them from growing beyond bounds - is equivalent to a high probability of the outbreak of war.* Many authors (e.g., Tuchman 1962, White 1985) have posited the loss of control as a frequently dominant aspect of the outbreak of war. It certainly seems important, then, both theoretically and practically (i.e., in policy making) to be able to distinguish between "safe" (stable) and "dangerous" (crisis unstable) ranges of values of the system parameters λ, so as to be able to steer away from the latter when appropriate.

Heuristically, I shall assume that the over-all-behavior of the world can be modeled by a system governed by deterministic laws such as Eq. (1), and that the mathematical structure of these laws is such as to allow the prediction of the onset of unpredictability. Using the language of Chapter 4, it may be very useful to be able to predict, and perhaps avoid, the occurance of chaos in the world system.

Thus, it seems useful, and in accordance with much previous writing (see Saperstein, 1990, for further discussion and references), to characterize war as chaos, to *assume as a working hypothesis that the transition from "cold war" to "hot war" in an international system can be usefully represented by the transition from non-chaotic to chaotic behavior in the non-linear model representing the interactions between the competing nations of the system.* If the onset of chaos is predicted in the model, then crisis instability and the outbreak of war is to be expected in the system being modeled. If the onset of war is to be expected according to a given model, and if such a war is not desired, then the model (or the model parameters) must be changed, i.e., the policy represented by that model must be altered.

As has been seen in Chapter 4, there exist simple determinate non-linear models, having the form of Eq. (1), that change from non-chaotic to chaotic as a model parameter is increased. For parameter values λ less than a critical value λ_C, the models have well-defined "attractors", making prediction quite easy. If the critical parameter value is exceeded, there is no end point; the path originating at any starting value will, in the course of time, range over the entire set of possibilities open to the system; any two neighboring paths will diverge, making prediction impossible. Such a chaotic system has a "strange attractor" - very different from the simple collections of points, smooth curves, or surfaces, which are the attractors of predictive systems. The important point is that the critical transition value can be determined within the model; *even though the model no longer allows the prediction of a specific evolution into the future, it does allow the prediction of this unpredictability.*

If such a non-linear model could "adequately" describe important aspects of the international system, it could then be used to predict crisis instability, i.e., the conditions under which the outbreak of war is likely.

In this regard, it is important to emphasize the difference between " strong chaos" and "weak chaos", first introduced in the previous chapter. In a model of international competition, strong chaos indicates complete uncertainty as to the outcome of the process, e.g., which nation will win if war breaks out, or indeed if the nation will survive the process. Weak chaos implies minor uncertainty, e.g., which particular soldier is killed in the inevitable march to "victory." It is the former which is of prime concern in this book.

To summarize the major paradigm of this book (Saperstein 1986), the presence or absence of chaos in a non-linear model implies stability or instability in the international system being modelled. This paradigm can then be used to answer such fundamental questions in international security theory as:

i. Are bipolar international systems more or less stable than multipolar international systems? (Saperstein 1991)

ii. Are democracies more or less prone to war than oligarchic nations? (Saperstein 1 992a)

iii. In an anarchic system, is the independent seeking of security via arms procurement more or less stable than the collective procurement of arms via balance-of-power politics? (Saperstein 1992b)

Each of the above questions will be re-expressed as: which of the two possibilities (bipolar vs multipolar, democracy vs oligarchy, independence vs balance-of-power) results in a greater region of stability, where by "region" is meant range of system parameter values?

A different simple model, of the form of Eq.(1), has been created to address each question but the method of analysis for each one is basically the same. The appropriate Liapunov Coefficients (which, according to Chapter 4, represent system instability - exponential separation of neighboring orbits - when they have a positive value) are computed, using simple spreadsheet routines on a desktop computer. The regions of system-parameter variation in which they are negative determine the relative regions of model stability; the greater the region, the more likely the "real system" is to be stable. It should be emphasized that no claim is made for the empirical validity of any of these models. They are presented as simple illustrations of the basic paradigm which, hopefully, contains some features of the international system essential to system behavior relative to the question at issue. More complete discussions and bibliographies are given in the references cited (Saperstein 1991, 1992 a,b).

3.Which Is More War Prone— A Bipolar Or a Tripolar World?

Is a bi-polar world more or less stable than a tri-polar world? We have left a world configuration dangerously dominated by two nuclear superpowers for a world perhaps dominated by none, one, or many. It is now important to gain insight as to whether we must be even more cautious than before or can be a little less timid in addressing many non-military security problems. One approach for obtaining an answer to this question is to compare the range of stability of a two power system with that of a corresponding three power world.

To specify a simple bipolar world, the general law of system evolution, Eq.(1), is *heuristically* taken to be the following simple pair of coupled coupled equations:

$$X_{n+1} = 4aY_n(I—Y_n) \qquad (2)$$

$$Y_{n+1} = 4bX_n(1 — X_n)$$

X_{n+1} , is the "devotion" of nation X to war in year n+1—the ratio of arms procurement (in a general sense, i.e., including societal infrastructure for war such as appropriate airports, harbors, highways, etc.) to GNP in that year, and thus lies between zero and one. In a first approximation, it is assumed proportional to the previous year's devotion of its opponent nation Y_n, and vice-versa (the more arms you buy this year, the more your opponent buys next year). But this linear co-variation can't go on forever; neither X nor Y may excede one. The non-linear term $(l\text{-}Y_n)$ enforces this constraint. It expresses the assumption that if the opponent's resources are "stretched to the breaking point," [i.e., his previous year's procurements were so great, so close to one, that they almost preempted his entire GNP (making it impossible for him to procure any more)], there is no need to stretch any further this year so that this year's procurements may be correspondingly diminished. Thus, when Y's armaments are small but increasing, so will X's (and visa-versa); when nation Y is reaching its limit, its opponent need not stretch itself unduly to increase its stocks. The system parameter λ, in this case, becomes the two constants "a" and "b", determining how the two competing nations react to each others threatening activities; they must also lie between zero and one so that X_n and Y_n remain between these bounds.

The region of stability of Eqs.(2) is analytically computed (Saperstein 1984) as a function of the model parameters a and b. The resultant curve represents the critical relation between a and b in the unit square of the a-b plane. The region above the curve, in which the two Liapunov coefficients $A(X_0)$ and $A(Y_0)$ are positive, is the model's chaotic region. Thus the square region in parameter space (0<a, b<l) is divided into crisis-stable and crisis-unstable regions for the system parameters a and b.

The model is now extended to three nations, X,Y, and X, in as simple and symmetric a manner as possible via the following trio of coupled equations (λ now consists of 4 numbers: a, b, ε ,c):

$$X_{n+1} = 4aY_n(I-Y_n) + 4\,\varepsilon\, Z_n\,(I-Z_n) \qquad (3)$$
$$Y_{n+1} = 4bX_n(I-X_n) + 4\,\varepsilon\, c\, Z_n\,(I-Z_n)$$
$$Z_{n+1} = 4\,\varepsilon\,[\,X_n(I-X_n) + c\,Y_n(I-Y_n)\,]$$

When ε (= epsilon) equals one, this three-nation system becomes the previous two-nation sysyem. As ε increases from zero, the coupling between X and Y and the third nation Z becomes increasingly significant. When ε = 1, the three nations interact completely symmetrically with each other. Numerical computations of the Liapunov Coefficients (Saperstein 1991), indicate that the stability region decreases in area as epsilon increases, i.e., as the third nation becomes more significant in the world system. Hence, the model suggests, in accord with a great deal of other scholarly evidence (e.g., Midlarsky 1989, Waltz 1979), that a tri-polar world is more dangerous than a bipolar one. (For a contrary view, see Deutsch and Singer, 1985.)

4. Are Democracies More or Less Prone to War?

Given the propensity of many of the world's major democracies to aid and abet less-than-democratic governments elsewhere, it would be useful to know if such policies hinder or further the peace that these democracies presumably seek and guard. To deal with this question, using the approach outlined in Section 2, we require first a definition of democracy which is easy to quantify and fit into a simple model of competitive arms procurement, such as Eq.(2).

It seems easiest to define democracy in terms of diffuseness of decision making. The larger the fraction of population which has significant input into matters of peace and war, the more democratic the nation will be considered to be. With this definition, the numerical difference between democracy and autocracy will be based upon population sampling theory. Given a pair of competing nations, designated by "X" and "Y", I assume that every citizen of nation X has some intrinsic feeling for the totality of its opponent nation Y, and conversely. These feelings may range from complete confidence in the other party to extreme fear and loathing. Let a_x be the "fear and loathing" coefficient for any one citizen of nation X. Then the policy of nation X towards its opponent will be determined by $<a_x>$, the mean of a_x over the class of citizens of X who are significant decision makers; $<a_y>$ is similarly defined. The fraction of citizens belonging to this class will be small in an autocracy, large in a democracy.

I assume, again heuristically, that there is a natural "spread" of values of the "fear and loathing" coefficient whose distribution is characteristic of humanity, not nationality. Each large nation will contain many people manifesting small values of the coefficient, intermediate values, etc. Thus the mean, â, of this coefficient, taken over a large population, is a human characteristic, independent of nation: $\hat{a}_x = \hat{a}_y = \hat{a}$.

The decision makers of a nation are a sample of the entire nation. If it is a very large sample, means over this sample will be close in value to means over the entire population. Hence, for large democracies, $<a> \cong \hat{a}$. Means over small samples, taken from the population, may vary significantly from the means over the entire population. Thus, given a random set of large autocratic nations, there will be some for which $<a>$ greatly exceeds â., some for which $<a>$ is much less than â.. In a corresponding set of large democratic nations, all will have intermediate values of $<a> = \hat{a}$.. Let a_{xy} be the *effective "fear and loathing" coefficient* of nation X for nation Y. If all nations in the set containing X and Y are large democracies, all will have the same "fear and loathing" coefficient, characteristic of the general human population: $a_{xy} = \hat{a}$. If the set contains autocracies, a_{xy} can range from very small values to very large values.This effective "fear and loathing" coefficient will be taken to be the proportionality constant in a non-linear arms procurement recursion relation similar to Eq.(2).

Denote the annual procurement of arms by X as ΔX (= $X_{n+1} - X_n$; note that in this section X_n is the total arms stock of X, not its annual procurement). In a competitive situation between nations X and Y, ΔX will be proportional to the amount by which X's arms stocks were exceeded by Y's in the previous year. The proportionality constant is a_{xy}; the similar coefficient for Y's annual arms build up is a_{yx}. Thus the first assumption is that $\Delta X = a_{xy} (Y_n - X_n)$, *if* Y_n is greater than X_n, and similarly for Y. X's procurement will also increase proportionally to the total strength of Y; if Y is very weak, it won't matter if Y is "stronger" than X; the more powerful Y is, the more damaging any discrepancy in power would be in any potential conflict. Thus this first assumption is altered to $\Delta X = a_{xy} Y_n (Y_n - X_n)$.

However, arms build-ups cannot continue indefinitely. There will be economic constraints; nations can't procure more than the total economy will allow. If C_x is the maximum annual expenditure for arms possible by X, the evolution equations must constrain ΔX to be less than C_x. A simple way to accomplish this is as follows: I introduce a continuous economic cut-off function

$$\tilde{\phi}(1 - \Delta X/C_x) = (1 - \Delta X/C_x) \cdot \Theta(1 - \Delta X/C_x) \qquad (4)$$

where Θ is the unit step function ($\Theta(x) = 1$ for $x > 1$, $\Theta(x) = 0$ otherwise). Thus $\tilde{\phi}(1 - \Delta X/C_x)$ falls smoothly from 1 to 0 as ΔX increases from 0 to its maximum allowed value C_x. The above assumptions can now be combined and written as $\Delta X = a_{xy} Y_n \tilde{\phi}(Y_n - X_n) \cdot \tilde{\phi}(1 - \Delta X/C_x)$, and similarly for ΔY.

If X's arms stocks exceed those of Y, then X can decrease its armaments. I assume that the arms build-down in any one year ("arms build-down" means that ΔX is negative) will be proportional to the amount by which X exceeds Y's strength in the previous year and also proportional to the total power of X: the more confident a nation is in its strength, the more it can afford to build down. The proportionality parameter for build-down, which will called the "confidence coefficient", is taken, for simplicity, to be the inverse of the build-up "fear and loathing" coefficient. Finally, it is assumed that "reconversion" of the "military-industrial complex" works so that there are no economic constraints for build down.

The result of combining the coupled arms build-ups and build-downs, for X and Y, is the following set of recursion relations, relating the total arms stocks of X and Y in any year (designated as "n+1") to the corresponding stocks in the previous year (n):

$$X_{n+1} = X_n + a_{xy} Y_n \cdot \tilde{\phi}(1 - [X_{n+1} - X_n]/C_x) \cdot \tilde{\phi}(Y_n - X_n) - (1/a_{xy}) X_n \cdot \tilde{\phi}(X_n - Y_n) \qquad (5a)$$

$$Y_{n+1} = Y_n + a_{yx} X_n \cdot \tilde{\phi}(1 - [Y_{n+1} - Y_n]/C_y) \cdot \tilde{\phi}(X_n - Y_n) - (1/a_{yx}) Y_n \cdot \tilde{\phi}(Y_n - X_n) \qquad (5b)$$

These relations may be iterated, given arbitrary starting values at $n = 0$, using standard spread sheet routines on a desk top computer. Going on to large n ($\cong 100$) (one can't get to infinity on a desk top computer!), the two Liapunov Coefficients can be calculated. Typical results are given in Figure 1, in which $a = a_{xy}$ and $b =$

a_{yx} . The region below the curve is the stable λ region: negative Liapunov Coefficients. Note that small values for both a_{xy} and a_{yx} imply stability whereas instability results when both a_{xy} and a_{yx} get large or when either gets very large.

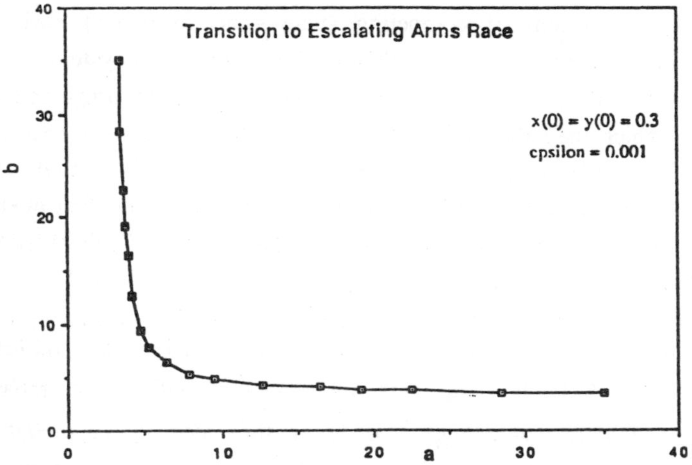

Fig.1. Stability plot for a bi-polar competetive system. The effective "fear-and-loathing coefficient" of one nation plotted against that of the other. The curve is the boundary between the region of stability below it, and the unstable region (crisis-unstable) above it. (From Saperstein, 1992a)

Since large values of the effective "fear and loathing" coefficient imply chaos and war, and since any collection of autocratic states (or large mixed collection of autocratic and democratic states) is more likely to have some pairs of nations with large values of this coefficient than a corresponding collection of democratic states, it follows that *an outbreak of war is more likely in a collection of autocratic states than in a similar collection of democratic states.* (For confirmation of this result via more traditional political science analysis, see, e.g., Russett, 1990.) Even though it is often easier in foreign affairs to deal with autocratic governments, it is evident that, in addition to the usual human desire to nurture similar systems elsewhere, democracies are more likely to ensure world peace by supporting the democratization of other members of the international system.

5. Which is More War Prone - a System of Shifting Alliances or a Collection of Go-it-Alone States?

As a further example of the paradigm of this book, illustrating the difference between the concepts of "strong" and "weak" chaos, consider the question: Which is a more stable system of competing states—one in which each nation individually procures arms so as to match any other state in the system; or one in which weaker nations join coalitions to collectively procure arms against the strong and in which they shift alliances when a coalition member becomes too strong—i.e., *independence security or balance of power security*? To model the two cases simply, consider three competing states, X,Y, and Z. In the first alternative, they will be independent of each other. In the second, any two will be allied against the stronger third.

Considering any pair of the three, arms build-up or build-down is presumed to occur just as in the previous case (Eq. 5) with a_{xy} the effective "fear and loathing" coefficient of X for Y and $1/a_{xy}$ the corresponding "confidence coefficient." However, now X may pair against Y or Z, in the *independence model* so the resultant arms procurement or build-down is the sum of the corresponding pair terms:

$$X_{n+1} = X_n + [a_{xy}Y_n \cdot\tilde{\phi}(Y_n - X_n) + a_{xz}Z_n \cdot\tilde{\phi}(Z_n - X_n)]\cdot\tilde{\phi}(1 - [X_{n+1} - X_n] /C_x) - (1/a_{xy})X_n\cdot\tilde{\phi}(X_n - Y_n) - (1/a_{xz})X_n\cdot\tilde{\phi}(X_n - Z_n) \tag{6a}$$

$$Y_{n+1} = Y_n + [a_{yx}X_n \cdot\tilde{\phi}(X_n - Y_n) + a_{yz}Z_n \cdot\tilde{\phi}(Z_n - Y_n)]\cdot\tilde{\phi}(1 - [Y_{n+1} - Y_n] /C_y) -(1/a_{yx})Y_n\cdot\tilde{\phi}(Y_n - X_n) - (1/a_{yz})Y_n\cdot\tilde{\phi}(Y_n - Z_n) \tag{6b}$$

$$Z_{n+1} = Z_n + [a_{zx}X_n \cdot\tilde{\phi}(X_n - Z_n) + a_{zy}Y_n \cdot\tilde{\phi}(Y_n - Z_n)]\cdot\tilde{\phi}(1 - [Z_{n+1} - Z_n] /C_z) - (1/a_{zx})Z_n\cdot\tilde{\phi}(Z_n - X_n) - (1/a_{zy})Z_n\cdot\tilde{\phi}(Z_n - Y_n) \tag{6c}$$

In the *alliance model*, X will join its procurement with Y to match a superior Z, or join with Z to match a superior Y, and similarly for build-down (here the appropriate coefficients are written b_{xy}, b_{xz}, etc.):

$$X_{n+1} = X_n + [b_{xy}Y_n \cdot\tilde{\phi}(Y_n - X_n - Z_n) + b_{xz}Z_n \cdot\tilde{\phi}(Z_n - X_n - Y_n)]\cdot\tilde{\phi}(1 - [X_{n+1} - X_n] /C_x) - (1/b_{xy})(X_n + Z_n) \cdot\tilde{\phi}(X_n + Z_n - Y_n) - (1/b_{xz})(X_n + Y_n)\cdot\tilde{\phi}(X_n + Y_n - Z_n) \tag{7a}$$

$$Y_{n+1} = Y_n + [b_{yx}X_n \cdot\tilde{\phi}(X_n - Y_n - Z_n) + b_{yz}Z_n \cdot\tilde{\phi}(Z_n - Y_n - X_n)]\cdot\tilde{\phi}(1 - [Y_{n+1} - Y_n] /C_y) - (1/b_{yx})(Y_n + Z_n)\cdot\tilde{\phi}(Y_n + Z_n - X_n) - (1/b_{yz})(Y_n + X_n)\cdot\tilde{\phi}(Y_n + X_n - Z_n) \tag{7b}$$

$$Z_{n+1} = Z_n + [b_{zx}X_n \cdot \tilde{\phi}(X_n - Z_n - Y_n) + b_{zy}Y_n \cdot \tilde{\phi}(Y_n - Z_n - X_n)] \cdot \tilde{\phi}(1 - [Z_{n+1} - Z_n]/C_z) - (1/b_{zx})(Z_n + Y_n) \cdot \tilde{\phi}(Z_n + Y_n - X_n) - (1/b_{zy})(Z_n + X_n) \cdot \tilde{\phi}(Z_n + X_n - Y_n) \quad (7c)$$

The two models each consist of three coupled recursion relations (Eqs. (6) for the independence model and (7) for the alliance model). The answer to the question posed is obtained by seeing which of the two models is more stable.

An illustrative simple analysis is done by assuming that the initial values X_o, Y_o, Z_o, for each of the three nations, are almost the same and asking whether these small differences grow as the recursion progresses. (Presumably the results would be quite different if the initial values were to differ from each other by some given finite amounts.) A spreadsheet iteration of the recursion relations is again carried out, starting off with X'_o, Y'_o, Z'_o, representing small random differences about an arbitrarily chosen starting point, X_o. Since there are now three variables, three Liapunov Coefficients are computed, varying the starting value X_o and the single recursion parameter a (representing a common value for a_{xy}, b_{xy}, a_{xz}, etc.).

Four possibilities become evident for the independent nation model as a result of the numerical computations (Saperstein 1 992b). In (I), "strong stability," all sequence limits exist and all three Liapunov Coefficients are negative, which implies that all sequence limits exist and are the same. There is complete predictability and hence no war. In "weak stability" (II), all sequences have limits, which are equal to each other, but the Liapunov Coefficients are positive. This occurs because all limits are less than the starting value X_o, except for the singular case when all start off exactly equal to X_o, in which case the limiting value is X_o. Hence the numerator in the definition of the Liapunov Coefficients doesn't become less than the denominator, implying positive coefficients. This situation is an illustration of"bifurcation" in that one point leads to results completely different from all other points. Since the limits are still well defined, the sequences are predictable, not chaotic, and so again there is no war. In the third case (III), "weak chaos", none of the sequence limits exist and so there are no Liapunov Coefficients. However, the variation within any one of the sequences X'_n, X'_n, X'_n, and the differences between the sequences $|X'_{n'} - X_n|$, remain small compared to the arbitrary joint starting value X_o The situation resembles case (b) in Figure 1 of Chapter 1, rather than case (c); hence, the unpredictability is minor and, in the paradigm of this book, does not represent war. The last possibility (IV) is one in which the sequences do not have limits and the variations within each sequence, as well as the differences between them, are large compared to the starting value. In this case the range of unpredictability is the entire range possible. There is no way of knowing - even approximately - the outcome of any policy or action, and hence major fluctuations may result from minor perturbations. This is the criterion for crisis instability and war.

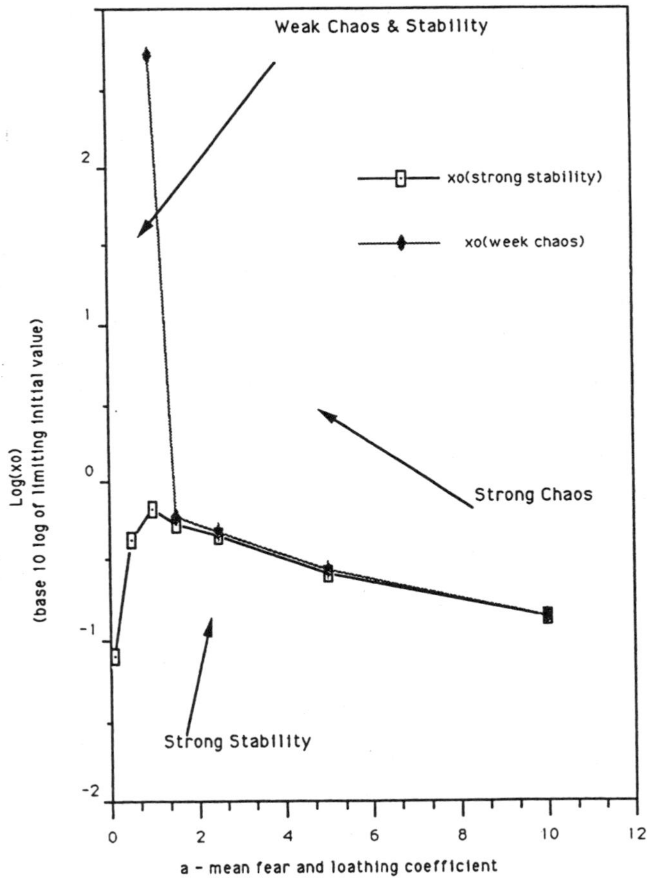

Fig.2. Stability plot for a model system of three independent competing nations. The region above and to the right of the curves is the region of "strong chaos" and presumed "crisis instability". (From Saperstein, 1992b)

The numerical results for the independent nation model are displayed in Figure 2. No sharp boundary was evident between the regimes of weak stability and weak chaos and so the two regimes are considered to be a single realm in which crisis stability is still valid. Note that there is no upper limit to the weak chaos regime, no breakdown of crisis instability, if the fear and loathing coefficient is small enough ($a \leq 1$). If there is enough confidence in the system ($1/a > 1$), perturbations to the independent nation international system may lead to continued fluctuations and uncertainty, but not to war no matter how large are the initial, equal, stocks of arms. On the other hand, with sufficient fear and loathing ($a > 1$), war will always break out if the initial arms stocks are large enough.

The numerical predictions of the alliance model are quite different. For large enough symmetric starting values, $X_0 = Y_0 = Z_0$, the system immediately and completely disarms no matter what value is chosen for the fear and loathing coefficient. This is region (I) of Figure 3; since all sequences tend immediately to a zero limiting value, it is a strong stability region. In region (II) of the figure, a limit has not yet been reached numerically at $n = 100$; none-the-less, the fluctuations in the monotonically decreasing sequences are small and thus this region is appropriately characterized as "weak chaos": the course of events is clear and the eventual complete disarmament is completely predictable.

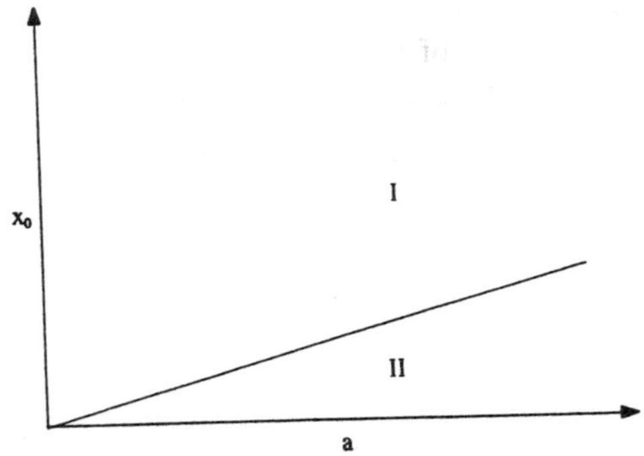

Fig.3. Stability as a function of mean starting value and mean "fear and loathing" coefficient for the balance of power security competition among three nations. In region I the system evolves immediately to disarmament. In region II, the system moves slowly towards disarmament in a weakly chaotic manner. (From Saperstein, 1992b)

Hence, using the possibility of strong chaos in the model as an indicator of the possibility of war in the corresponding international system, it follows that *a policy of building (perhaps shifting) alliances is guaranteed to keep an initially symmetric system peaceful. A policy of seeking national security by "going it alone" allows the possibility of war in such a system if the contending parties start out with large enough levels of armed force.*

I know of no "real world" situation in which a system of comperable competing states has balanced" itself down to complete disarmament, as predicted by the alliance model of this chapter. Perhaps this is because a shifting of alliances has costs in the real world, costs not included in this model. Also possible is that there are no real balance-of-power situations; real nations may resort to balance-of-threat instead, a much harder situation to model (Walt 1987). 0r, the presumed anarchy may not have always been present.

6. Conclusions

Much of the "traditional" approach to addressing questions of international security (or any other political or human interaction question) is based upon "intuition" - a compilation of the failures and successes of previous interactions with similar questions. The political world has been changing very rapidly in the last few generations: from a world of large distances between major international "actors", and big time intervals between events, to one of close contact and possibly immediate response; from one of relatively weak influence upon the domestic situation from far-away events to one of perhaps shattering or obliterating impact; from one in which we may respond to another's actions to one in which we must account for the other's response to our own actions when we decide what to do; *from a linear world to a non-linear world.* Our intuitions are still generally linear intuitions and may not be the most effective for dealing with present and future challenges to international security. (The usual notions of"national sovereignty" are a good example of linear intuition.) It is important to acquire a "non-linear intuition" (Prigogine and Stengers 1984) but perhaps dangerous to acquire it from possible actual nuclear war experiences. Thus, non-linear numerical simulations such as those illustrated in this chapter may be more valuable than the specific results obtained. (The military has long recognized the value of simulations - "war games" - as a preparation for dealing with "reality".)

There are many obvious situations which can be viewed as aspects of strongly interactive (i.e., non-linear) systems and hence be modelled accordingly. Among them are:

a). More realistic representations of the three questions already addressed in this paper, e.g., including the "costs" of alliance formation and dissolution.

b). S.D.I. - the "Strategic Defence Initiative of the Reagan Administration ("Star Wars") or other aspects of strategic defense and offense (Saperstein and Mayer-Kress 1988).

c). Battle and attrition models (going beyond the simple Lancaster-type) that explore the potentialities of "non-provocative" defense in a highly intertwined, technological world of vastly disparate powers (Saperstein 1986b, 1987).

d). Successes or failures in the verification of "conventional", nonnuclear "unconventional", as well as nuclear arms control agreements. Such agreements are only likely to be negotiated and adopted if there is some understanding of the future role of verification of non-strategic, as well as strategic, capabilities and readiness.

Non-linearity will be important in the consideration of all of these issues though the role of chaos may vary significantly. Much remains to be done in the generation and dissemination of non-linear intuitions for the policy makers of future generations.

7. References and Additional Readings

Deutsch, Karl W. and J. David Singer. 1985. In *Internatiotal War,* edited by Melvin Small and J. David Singer, Homewood, IL: Dorsey Press. Reprint of "Multipolar Systems and International Stability," *World Politics*, **16** (April): 390-406.

Farrar, L.L. Jr. 1985. *In International War,* edited by Melvin Small and J. David Singer, Homewood, IL: Dorsey Press. Reprint of "The Limits of Choice: July 1914 Reconsidered," *Journal of Conflict Resolution,* **16** (March): 1-23.

Kennedy, Paul 1987. *The Rise and Fall of the Great Powers.* New York: Random House.

Midlarsky, Manus I.1989. "Hierarchical Equilibria and the Lon-run Instability of Multipolar Systems". In *Handbook of War Studies,* edited by Manus I. Midlarsky, 55-81. Boston: Unwin Hyman

Prigogine, I. and I. Slengers 1984. *Order Out of Chaos.* New York: Bantam.

Reagan, President Ronald 1983. "Korean Airline Massacre", Washington, D.C.: U.S. Dept. of State, Bureau of Public Affairs, Office of Public Communication, Editorial-Sept. 5, 1983.

Russett, B. 1990. *Controlling the Sword* . Cambridge, Mass.: Harvard University Press.

Saperstein, Alvin M. 1984. "Chaos - A Model for the Outbreak of War,"*Nature,* **309**: 303-305.

Saperstein, Alvin M. 1986a. "Predictability, Chaos, and the Transition to War," *Bulletin Peace Proposals,* **17**(1): 87-93.

Saperstein, Alvin M. 1986b. "Fluid Dynamic and Kinetic Theory Models for a Non-Provocative Land Defense of Central Europe," *American Journal of Phvsics,* **54**(7): 607 - 611.

Saperstein, Alvin M. 1987. "Attrition of Aggressive Armored Forces by Local Militias in a Depletion Zone - An Enhanced Non-Provocative Defense in Europe," *Journal of Peace Research,* **24** (1): 47 - 60.

Saperstein, Alvin M. 1990. "Chaos and the Making of International Security ç Policy," in *The Ubiquity of Chaos,* edited by Saul Krasner, Washington, D.C.: American Association for the Advancement of Science.

Saperstein, Alvin M. 1991. "The Long Peace - Result of a Bi-Polar Competitive World?,"*Journal of Conflict Resolution,* **35** (1): 68-79.

Saperstein, Alvin M. 1992a. "Are Democracies More or Less Prone to War? A Dynamical Model Approach," *Mathematical and Computer Modelling,* **16**(8/9): 213-221.

Saperstein, Alvin M. 1992b. "Alliance Building vs. Independent Action: A Non-Linear Modeling Approach to Comparative International Stability," *Journal of Conflict Resolution,* **36** (No. 3): 518-545.

Saperstein, Alvin M. 1994. "Chaos as a Tool for Exploring Questions of International Security", *Conflict Management and Peace Science,* **13,** no. 2, 149-177.

Saperstein, Alvin M. and Gottfried Mayer-Kress. 1988. "A Non-Linear Dynamical Model of the Impact of S.D.I. on the Arms Race," *Journal of Conflict Resolution,* **32**(4): 636 - 670.

Schrodt, Philip A. 1981. *Preserving Arms Distributions in a Multi-Polar World.* Denver: University of Denver Monograph Series in World Affairs, Volume 18, Book 4.

Tuchman, Barbara. 1962. *The Guns of August.* New York: Dell.

Walt, Stephen M. 1987. *The Origins ofAlliances.* Ithaca, New York: Cornell University Press.

Waltz, Kenneth N. 1979. *Theory of International Politics.* Reading, MA: Addison-Wesley.

Weart, Spencer R. 1998. *Never At War: Why Democracies Will Not Fight One Another,* New Haven: Yale University Press.

White, Ralph K. 1985. In *International War,* edited by Melvin Small and J.David Singer, Homewood, IL: Dorsey Press. Excerpt from *Nobody Wanted War.* 1968. Garden City: Doubleday, pp 7-20 on the onset of WWI.

CHAPTER 6

COMBINED MODELS: CAPABILITIES AND INTENTIONS

1. Capability vs Intent

In the previous chapter, the paradigm of this book was applied to the problem of the initiation of war using models based upon the interactions between the military capabilities of the competing states in the international system. As a result of these interactions, the *capabilities* changed with the evolution of time and the system. Whenever the *intents* of the nations towards each other entered the model ("fear and loathing", "confidence"), they entered just as fixed model parameters. But *intents do change with the evolution of the real international system: perceived capabilities modify intents, and intents lead to changes in capabilities.* In this chapter an attempt is made to model this interaction between capability and intent. Again, the goal is to illustrate method, and to develop insight and intuition, rather than to achieve realistic results. Again, the procedure is carried out by addressing the same important questions of current interest to International Security theorists.

(1) What is the role of "arms races" between competing nations in an anarchic world (Waltz, 1979) in ensuring or undermining the security of that world?

(2) How do alliances (Walt, 1987) or enmities between competing nations evolve in such a world?

Two different types of mathematical-dynamical models of the world system have been commonly used in addressing these questions. For the first question, as was seen in previous chapters, Richardson (1960a,b) and followers (e.g. Saaty 1968; Schrodt 1981) have considered the *military capabilities* of competing states as the sole determinants of the "arms race" and have written linear interactive equations for the time evolution of these capabilities. As has been seen, Richardson models the competitive process, but there is no "war" in his equations or their solutions. My nonlinear Richardson models, in which crisis instability in the international system is associated with the transition from predictability to non-predicatability in the mathematical model, still involve only capabilities as system variables.

Lee, Muncaster, and Zinnes ("LMZ"; cf., Chapter 2) have approached the second question using the alternative type model: their dependent variables are the "relationships" between pairs of states, their mutual "feeling" of friendship or hostility, hence their *intent toward each other.* They postulate a set of first-order (in time) non-linear differential equations which model the presumed "structural rules" of a competitive international system - rules which govern the variation of *intent..* The resultant evolution of intent, according to these equations, depends solely upon the prior values of these mutual intents. Their interest is in the time evolution of specific solutions rather than their differentiation into stability classes.

Common sense suggests that *capability and intent influence each other.* Thus, such a relationship should be included in any "realistic" model of an arms race in the current world system. Given any pair of nations in such a system, awareness, by one nation, of the military capability of the other, should have a strong influence (but *not* the sole influence) upon the intent of the one towards the other. (Increasing military power of one of your competitive "partners" in the international system should make you at least feel "uneasy" about that partner, though the unease might be mitigated by other factors.) Similarly, one partner's intent towards the other, and its inference of the other's intent towards itself, would be expected to be a major influence in the evolution of the first partner's military capability. (If you consider yourself to be among enemies in the world system, or to be an enemy to others in the system, you are more likely to attempt to attempt to strengthen yourself militarily.)

The purpose of this chapter is to continue the demonstration of how mathematical international security models are created *and interpreted* by creating mathematical models based upon the approaches of both Lee, Muncaster, and Zinnes and of Richardson. These models will include both relationship *and* capability and the influence of each upon the other. They will be non-linear and thus capable of exhibiting a change of state from predictability to chaos. It should, therefore, using the paradigm of this book, now be possible to represent the transition from peace to war (crisis instability) in the corresponding world system. Thus, by varying the parameters of the models it may be possible to obtain a qualitative "feel" for the corresponding factors' influence upon the occurrence of war or peace in the international system.

I emphasize that the models in this book are, at best, incomplete. At this stage in the research program no attempt has been made to fit the model parameters to empirical data, nor to claim that the models provide recognizeable "pictures" of the real international system, past or present. What is attempted is the creation of the political science analog of the orrery - a mechanical model of our solar system which provides a representation of *some* of the important phenomena of prime interest at the moment. (An orrery might be set up to successfully represent the phases of the planets without being able to represent the scale or dynamics of the solar system. A different orrery may be able to represent the distance scales without being able to demonstrate the planetary phases.) The hope is that these simple models will indicate the relevant factors, and their interrelationships, which will be important to the eventual understanding of the questions raised at the begining of this chapter.

The result of a qualitative analysis of the simple model of this chapter (independent of the values of the system parameters) is that neither pure military capability nor pure mutual intent necessarily lead to crisis instability and war. In order for the model to be able to represent the possibility of crisis-instability and the change-of-state characteristic of the transition from peace to war, *it must include both military capability and relationship*, and the interplay between these two very different theoretical concepts. Neither the pure Richardson nor LMZ models, alone, can *explain* the outbreak of war in a three-nation system; however a combination of the two theories *allows* such an explanation of the outbreak of war.

Whether or not the change-of-state *necessarily* occurs depends upon the specific values of the model parameters. That question has not been pursued here since it has not been possible to assign "general" values to the many model parameters. (Determination of reasonable sets of such values would be a major research program, perhaps to be productively pursued in the future by readers of this book.) However, *sufficient conditions for the absence of war in the combined, extended, model can be found: war will not occur if "iron triangles" (i.e military-industrial-governmental complexes) have little influence upon the evolution of overall economic-political powers and/or if changes in military capabilities do not lead to changes in national intent towards others.* The influence of "iron triangles" on the propensity towards war is often noted (e.g., Kennedy 1987). Also often suggested is the role of "turning of the other cheek" to avoid conflict escalation; e.g., don't change your attitudes just because of a display of strength by the other party. Thus, as expected, the mathematical approach of this book leads to policy implications which concur with a great deal of folk and scholarly wisdom.

2. Modification of the Richardson Model to Include Inter-nation Intents and Economic Constraints

The two-nation Richardson arms race model of Chapter 3 is easily generalyzed to the following N-nation model:

$$\dot{x}_i = \Sigma_j a_{ij} x_j + f_i \tag{1}$$

where $x_i = x_i(t)$ is the positive-valued time-dependent variable representing the Richardson military capability of nation i; the dot over the variable, as before, represents the time rate of change of that variable, i.e., it's time derivative (i = 1 N, for an N nation system; for N =2, x_1 = x of Chapter 3 and x_2 = y). The sum, Σ_j, (j = 1,...N) is over all of the N nations of the system. For i ≠j, the N•(N- 1) system parameters a_{ij} represent the inate abilities of nation i to acquire arms in response to the arms of nation j (the "distrust and suspicion" constants "a" and "b" of Eq.1 of Chapter 3). The N system parameters a_{jj} ,where i=j =1,...,N, are the "military complacencies" of each of the N nations (the "c" and "d" of Chapter 3). The N numbers f_i are the system *constants* representing response of nation i to the fixed intents of the remainder of the system ("e" and "f" of Chapter 3). Everything that was said about Chapter 3's Eq.1 and its solutions is equally valid for the above Eq.1.*The goal of this chapter is to make each* of the intent parameters a variable *and to make each of the capability parameters depend upon the intents.*

As we have seen in Chapter 2, Lee, Muncaster, and Zinnes (LMZ) introduced a variable r_{ij} to represent the relation between nations i and j. A positive relationship, $r_{ij} > 0$, represents friendship or cooperation between i and j; a negative relationship, $r_{ij} < 0$, represents enmity or conflict:. (The variables r_{ii} are identically zero for each nation since national introspection is not under consideration.) Thus, the most obvious modification to make in the original Richardson model of Equation (1) is to assume that the original intent constants are now intent variables, proportional to r_{ij}.

$$f_i = - \Sigma_j r_{ij}\gamma_j \tag{2}$$

where the γ_j are N positive parameters representing the parcelling out of Richardson's innate trust or hostility among all of the N nations of the international system. The negative sign is required since positive relationships imply trust, implying a build-down, a negative time derivative, of military capability.

The next modification is to meet the requirement that a nation's response to the buildup of arms by another state (as represented by the a_{ij} term) must be conditioned by the total political-economic power of the nation. A nation must have the resources, and the will and ability to marshall them, before it can build military capability. Let p_i (the N "economic parameters") be some measure of the total national resources of country i. and its ability to deploy them. (At some minimal level, ignoring the question of political will to marshall and deploy resources, the annual rate of change of p_i would roughly be the GNP of nation i). A simple linear relation between a_{ij} and p_i is unreasonable since nation i will not be able to acquire any military capability at all, no matter what the perceived threat from j, if p_i is too small (there are mouths in i to feed first). The simplest analytical way to approximate a threshold relationship is to assume a quadratic relation between i's capability-response to capability and its economic capacity:

$$a_{ij} \rightarrow (p_i)^2 a_{ij} \tag{3}$$

With few resources available, a nation's response to another nation's armaments can only be minimal; as p_i increases beyond a threshold ($p_i \sim 1$), major military capability responses become possible.

Finally, a nation's response to another's arms buildup must depend upon the intent between the two nations. A hostile relationship is certainly going to call forth a similar buildup whereas a friendly relationship implies less of a buildup or even a build-down, the latter if there is a dependency upon the allies' military capability for protection. Hence, the response coefficients a_{ij} must be modified to depend upon the relationship variables. The simplest modification is:

$$a_{ij} \rightarrow a_{ij} (1 - \Gamma_j r_{ij}) \tag{4}$$

where the Γ_j are N positive parameters. A negative "relationship" between i and j implies a larger response by i to j's buildup than does a positive relationship. Since $r_{ii} = 0$ there is no modification to the "complacency" term (i=j). Thus, the suggested "reality" revised Richardson model is:

$$\dot{x}_i = \Sigma_j (p_i)^2 a_{ij} (1 - \Gamma_j r_{ij}) x_j - \Sigma_j r_{ij} \gamma_j \tag{5}$$

where the Γ and γ coefficients are measures of the impact of relationship upon the evolution of military capability. (Since they represent the conversion of intent into "action" - changing military capability, they will be referred to as "action parameters"). Equation (5) is a simple modification of the original Richardson model to explicitly include intent and economics as factors in the time evolution of military capability.

If the quantities p_i and r_{ij} are taken to be fixed parameters or functions of time determined exogenously to the model, the resultant set of N equations, Equation

(5), have analytical properties similar to that of the unmodified Richardson equations, Equation (1). They would still be linear, first-order, coupled differential equations, hence integrable and thereby non-chaotic. Though the solutions may no longer be the simple exponentials of Chapter 3 (if the economic and action parameters are time dependent), they would not manifest the possibility of a "change of state" with change of parameters. Such a model still would not represent a crisis instability in the international system and the possible transition from peace to war. But economic constraints and intents should themselves be time-dependent variables in any "realistic" model. In the next sections, $p_i(t)$ and $r_{ij}(t)$ will be determined by a suitably extended model, making Equation (5) non-linear with the possibility of solutions of a radically different character.

3. Modification of the LMZ Model of Intent Variation to Include Military Capability

As seen in Chapter 2, the LMZ model is based upon the "structural rules" of international anarchy, which govern the time evolution of the "relationships" between states (Eq.10 in Chapter 2). In its simplest form it can be written as:

$$\dot{r}_{ij} = Ar_{ij} + B\Sigma_k r_{ik}r_{kj} \tag{6}$$

where A and B are fixed positive parameters (we have assumed that the A and B of Chapter 2, Eq. 10, are the same for all nations, hence independent of the variables i,j,k). The linear term A represents the "direct structural interaction rule": "Any friendship of country i towards country j is enhanced by immediate past friendship and any conflict is enhanced by immediate past conflict". The non-linear term B, with its sum over all third nations k ($k \neq i$, $k \neq j$), represents the "inductive" and "comparative" "structural interaction" rules: "Any enemy of my enemy is my friend, . . .". If i and j are both positive to a third nation k, or both negative to k, the relation between i and j improves and r_{ij} increases. Conversely, if i is hostile to one of j's friends k ($r_{ij} < 0$, $r_{kj} > 0$), or if the relation between i and j deteriorates, then $\dot{r}_{ij}$ decreases. In its simple form, i.e., Equation (6), the model assumes a symmetric relationship between i and j so that $r_{ij} = r_{ji}$, for all time (i's intent towards j is identical to j's intent towards i).

A number of interesting theorems, concerning the time evolution of alliances and antagonisms from given initial configurations of positive and negative relationships, have been derived for the LMZ model. However, at least for the three-nation system, N = 3, these evolutions are purely predictive - they are non chaotic; hence they cannot undergo the mathematical change-of-state which, in the paradigm of this book, is to be taken as a theoretical represention of a real peace-war transition.

In the model given by Equation (6) the evolution of the relationship between the nations is driven solely by existing relationships; national military capabilities have no influence which does not seem reasonable. Enmity is expected to be enhanced beyond that merely due to past conflict if the enemy's military capability is increasing and decreased if its military power is decreasing. Similarly, if your friend is building up his arms, your increasing ardor towards him might be dampened somewhat by a hesitant suspicion and vice versa. (In the original Richardson model, any build up of arms by any other nation leads to a build up in response, certainly not a sign of increased friendship). Hence, a reasonable, simple, modification to the "direct structural interaction rule" is:

$$\dot{r}_{ij} = Ar_{ij}\,(1 - \alpha_{ij}\, r_{ij}\, \dot{x}_j) \tag{7}$$

where the positive coefficients α_{ij} are measures of the impact of military capability upon the evolution of inter-nation intent. If i and j have a negative relationship, this relation deteriorates faster if j is increasing rather than decreasing its armaments. The variation is reversed if the initial relationship is positive.

The effect of mutual intent and capability upon inter-country relationships should be more complex when more than two nations are involved. "Suppose further that i and j are not only friendly but also, jointly, very hostile to k. Then you could argue that their combined positive rates (of increase of military capability) actually enhances their mutual friendship." (Zinnes, 1992). This additional, three party effect, commonly called "load-sharing", can be represented by modifying the non-linear term of Equation (6) to

$$B\, \Sigma_j\, r_{ik}\, (1 - \beta_{ik}\, \dot{x}_k\, r_{ik})\, r_{kj}\, (1 - \beta_{kj}\, \dot{x}_j\, r_{kj}) \tag{8}$$

where the β_{ij} are positive "capability-intent" coefficients. This mathematical expression represents "load-sharing", "band-wagoning" (Walt 1987), and the Richardsonian suspicion of any arms build up on the part of any other nation. These are discussed in turn.

(a) "Load sharing" describes the following possibilities, all implicit in Eq.8, if k has a negative relationship with both i and j: If the rate of increase of the i - j relationship, $\dot{r}_{ij}$, is enhanced when both k and j have positive arms buildups, i.e., I increase my ties to my friend who is building up in the face of my enemy. If k and j both build-down, i has less need of a less powerful friend j in confronting a less powerful opponent k and hence diminishes the rate of increase of its relationship to j. If, on the other hand, j builds up while k builds down, the rate of change of i's relationship to j can be either enhanced or diminished depending upon the specific values of the constants and variables; enhancement represents a ganging up on a weakening opponent; diminishment implies less need by i of j's help. Conversely, if j builds down while k builds up, enhancement and diminishment of the i - j relationship are again both possible; diminishment implies a resentment by i that j is not sharing the load; enhancement implies that i may need j's help. Finally, if both j and k builddown, i's relationship with j grows at a less rapid rate since i has less need of j.

(b) "Band-wagoning" is said to occur if a nation leaves its existing alliances to join with an alliance which is obviously growing stronger. It manifests itself in Eq.8 as follows when k has a negative relationship with i, a positive one with j: If both j and k engage in arms buildup, the decline in the i-j relationship can either be accelerated - both my enemy and the friend of my enemy are getting stronger so I'd better become more hostile to the latter, or abated - if both are getting stronger, perhaps I'd better lessen my enmity and join the bandwagon. If j and k both build down, again the rate of decline in the i - j relationship can be either enhanced or diminished, depending upon the parameter and variable values; diminished hostility growth could be a consequence of everybody's disarming; enhanced hostility growth could be a sign of disrespect for a weakening opponent. If k builds up while j builds down, i's negativity towards j is again enhanced, again a sign of disrespect for "wimps". If, conversely, k builds down while j builds up, the rate of enmity increase between i and j is diminished which may represent joining the bandwagon of the strengthening opponent.

(c) "Richardsonian suspicion" is said to occur if a nation responds to increasing/decreasing armaments on the part of any other nation by building up/down its own military capability, a feature of the original Richardson model still maintained in Eq.8. If k has positive relationships with both i and j while both k and j engage in an arms buildup, the positive i - j relationship will grow at a diminished rate - a manifestation of the Richardsonian suspicion of anybody else's arms buildup. If k and j both decrease their armaments, the i - j relationship is enhanced; since everybody is cutting their arms, i can feel even better towards its friends. If j builds up while k builds down, or vice-versa, the growth of i's relationship to j may be enhanced or diminished depending upon the specific values; diminishment would again represent Richardson's suspicion of some arms buildup; enhancement could represent Richardson's complacency at arms build down.

The resulting modified LMZ model, which incorporates the rate of change of military capability upon the time evolution of internation intent, can now be written as

$$\dot{r}_{ij} = Ar_{ij}\,(1 - \alpha_{ij}\, r_{ij}\, \dot{x}_j\,) + B\,\Sigma_j\, r_{ik}\,(1 - \beta_{ik}\, \dot{x}_k\, r_{ik})\, r_{kj}\,(1 - \beta_{kj}\, \dot{x}_j\, r_{kj}) \qquad (9a)$$

This form does not posses the mathematically desirable symmetry expressed by $r_{ij} = r_{ji}$. The desirable symmetry is also more politically "realistic" — it does not seem likely that i's intent towards j would long be markedly different from j's intent towards i, assuming that all intents become known to all. To retain the desired symmetry, the model is modified to

$$\dot{r}_{ij} = Ar_{ij}\,(1 - \alpha_{ij}\, r_{ij}\, [\dot{x}_i + \dot{x}_j\,]) + B\,\Sigma_j\, r_{ik}\,(1 - \beta_{ik}\, \dot{r}_{ik}\, [\dot{x}_i+\dot{x}_k])\, r_{kj}\,(1 - \beta_{kj}\, r_{kj}[\dot{x}_k+\dot{x}_j]) \qquad (9b)$$

Model Equation (9b) represents essentially the same structural rules as the model of Equation (9a). Since α and β determine the impact of changing military capabilities upon internation intent, they can also be referred to as "verification parameters". Note that a large enough rate of change of military capability can change the sign of the factors in Equations (9b). Thus, sufficiently rapid alterations in military capabilities can change growing friendships into animosities, or vice versa. These modifications are reasonable first approximations to the explicit impact of capability upon intent. Note also that the model (differential) equations are now considerably more complex than originally, whether the time rates-of-change of military capability, $\dot{x}$, are given parameters or model determined via some form of Richardson Equations. If the military capabilities are given functions of time, the equations can be solved (numerically, if no other way) to explicitly show how an original configuration of international relationships evolves under the impact of perceived changes it military capabilities. Or, without solving the equations, the qualitative behavior of their solutions - their stability - can be examined. Given that capability changes influence intents, the desired system transformation into chaos may very well be possible in this modified LMZ model. But first the effect of the variation of economic capacity (the "economic parameters") should be examined, as they constrain the arms race via Eqs. (3) and (5).

4. A Simple Model for the Impact of Intent and Military Capability Upon Economic Capability

Resources for building up military capability can only be obtained fror the overall economic and political resources of the society. This overall capability in turn depends upon the military capabilities extent in the world system. For example, it is well known that foreign threats or "enemies" can bring forth sacrifice and social cohesiveness and consequent economic-political growth. Hence an arms race model should include a relation between the variables $p_i(t)$, $x_i(t)$, and $r_{ij}(t)$. The simplest abstract societal model for the time evolution of overall capability p_i (ignoring the physical-biological constraints of a finite earth) is:

$$\dot{p}_i = c_i\, p_i + \eta_i\, \Sigma_j\, r_{ij}\, p_j + \Sigma_j\, \sigma_{ij}\, x_j \tag{10}$$

The first term of (10), with positive coefficients c_i, is the simplest representation of the assumption that national wealth p_i leads to the creation of further wealth $\dot{p}_i$. The second term, with positive coefficients η_i (also "action parameters"), implies that the friendly intent between two nations will lead to a growth in wealth via trade and other forms of social/technological cooperation whereas a hostile relationship may lead to the diminishment of wealth via negative trade practices, hindering the flow of necessary resources, etc. In both cases, the impact of the intent towards each of the second nations, j, is assumed to be proportional to the size of that second nation's overall wealth p_j.

The signs of the third set of coefficients σ_{ij} are not so obvious. There is much disagreement as to whether a strong military capability of a nation (a large "military-industrial-complex" or "iron-triangle") leads to an increase in overall societal capability of that (i=j) nation ($\sigma_{ii} > 0$) or a decrease ($\sigma_{ii} < 0$). (The former was clearly the case for the British Empire, and the latter was clearly the case for the Soviet Union). Similarly, the impact of another (i≠j) nation's armaments upon the total wealth of a given nation is not generally well established. Hence the signs of the σ_{ij}, referred to as the "iron-triangle parameters", will be left unspecified. It is also possible that the impact of the nation j's armaments upon i depends upon the relationship between i and j, in which case the third term of Equation (10) should be multiplied by r_{ij}. (If this is done, then the diagonal term must be separately inserted since $r_{ij} = 0$ for i=j.)

Note that the N Equations (10) are just simple, linear, first-order differential equations, easily integrable, if the time dependent variables $r_{ij}(t)$ and $x_i(t)$ are parameters specified from outside of the model. In this case the solutions are predictable with no possibility of chaos or crisis instability. If the r_{ij} and x_i are constants, the solutions to Equations (10) are again simple exponentials — oscillating, rising, or falling. As such, the model Equation (10) is a simplified version of standard linear input/output matrix models such as that of the "Club of Rome" (Meadows, et al. 1972).

5. Qualitative Consequences of the Combined Model

A more realistic model of an "arms racing" international system should include the overall (p) as well as the military capabilities (x) of each nation and the relationships between the nations (r) as dependent variables as well as the influence of these factors upon each other. Such an extended model could be represented by the coupled set of differential equations (5), (9), and (10). Instead of each being a separate model, determining its own single dependent variable, x, r, or p, with the other two treated as given independent parameters, all three variables are to be dependent variables in all three equations, conjointly determined simultaneously by the three sets of equations. In this generalized model, parameter sets Γ and γ represent the impact of mutual intent upon military capability, η its effect upon overall capability, α and β the impact of military capability upon mutual intent, and σ its effect upon overall capability. In addition to the six sets of Greek letter parameters which couple the three sub-models (5), (9) and (10) to each other, there are also the four Latin letter parameters A, B, a, and c, exclusively within the sub-models. Thus, a total of ten parameter sets are required to completely specify the model.

The combined model is presumed to be a better representation of the behavior of the world system than the previous models employed in this book, though it is still, at best, only a partial representation of the system. Its parameters, as has been discussed in previous chapters, are determined by national characteristics — economic, historic, physical, social, etc., or by deliberate policy choice. Presumably, if all of these characteristics could be unambiguously and quantitatively expressed, the model parameters would be uniquely determined, and hence the time evolution of the model. However, as has previously been noted, these domestic characteristics are themselves strongly influenced by the evolving world system — hence the parameters representing them are not expected to remain constant. Nevertheless, the rate of variation of the non-policy parameters should be much less than that of the international system; national *socio-economic* characteristics are assumed to vary on a much longer time scale than international *political* characteristics. Hence, in a given "epoch", the system parameters are assumed to be constants which determine the evolution of the arms race aspects of the international system via Equations (5), (9), and (10). (The variation of the parameters, and the progression from "epoch" to "epoch", would - in principle - be determined by a much more complete, and complex, model governing the internal evolution of societies and of their individual members.)

The complete combined model is much more complex than its three component sub-models, which separate out from it when all of the Greek letter parameters are set equal to zero. *Solutions to the combined model may exhibit chaotic and non-chaotic regions and the possibility of transition between them as the parameters are varied. This is a necessary characteristic of any model capable of representing that aspect of the international system which undergoes peace-war changes of state.* Though the combined model developed here does not realistically represent the entire international system, its behavior should sufficiently mimic the arms race aspects of the real system so as to provide some guidance for policy making in that realm over the short- and medium-range future. In such a model, the parameter value ranges which allow chaotic solutions are the dangerous values, the values which should be avoided by statesmen hoping to avoid war in the international system.

If specific parameter values could be obtained from "real world data", these values could be substituted into the model equations. The models could then be examined for the possibility of chaos, e.g., by computing their Liapunov exponents. However, a major research program will be needed, comparing model and empirical world behavior over extended time periods, so as to refine the functional forms assumed for the model and to determine the plausible ranges of the parameter values. Only then will the model allow such immediate comparisons. It will not be possible, for a long time, for a theorist to say to a statesman that a particular national policy, in association with the known national characteristics, will lead the nation into danger.

Barring such specific predictions, quantitive models might be used for the suggestion of general trends of behavior with the variation of some parameters, the answering of some general theoretical questions (e.g., the discussion in Chapter 5). However, there are too many free parameters in the combined model of this chapter to allow a search for "interesting" values without the empirical guidance of the research program mentioned above. The dimension of the parameter space is just too large to allow a random search for interesting regions. It follows that all that can be done at present is to seek *necessary or sufficient conditions for the possibility or impossibility of chaos* = crisis instability and the possibility of war in the model. Hopefully, inasmuch as the model does reflect some important aspect of reality, this will lead to some interesting insights into the "real world". Such "seeking" is done in the remaining paragraphs of this chapter.

If the parameters α and β (the "military capability→intent" parameters) are all zero, the intent variables r evolve independently of the world system's military and overall capabilities. Then, as shown by LMZ (cf., Chapter 2), for a three nation system, the relationships evolve smoothly and predictably, i.e., there are no chaotic transitions. If, in addition, the σ ("military capability→total capability" parameters) all vanish, the overall capabilities evolve independently of the military capabilities. (Since the r(t) would have already have been determined, the equations for $\dot{p}$ would be linear and, thereby, also non-chaotic.) Then r(t) and p(t) would be pre-determined functions in the $\dot{x}$ equation which itself would also be linear and hence incapable of manifesting a chaotic change of state. Hence, a *sufficient condition for the impossibility of chaos in the combined model* is that $\alpha = \beta = \sigma = 0$. This represents a world with no "iron triangles", in which changes of military capability do not lead to changes in national intents towards other nations. Such a world will not succumb to war, even in the midst of an arms race. Conversely, in an arms-racing world, the presence of one or more "iron triangles", which enable changing military capabilities to influence international relationships, *is a necessary — though not sufficient condition for the existence of crisis instability and hence the probable outbreak of war.* There are, presumably, ranges for these variables which lead to sufficiency and war and other values which make war impossible, but these ranges could only be determined once values for all of the other model parameters were specified.

The above sufficiency requirement seems interesting, if not particularly new. Assuming, as Richardson did, that war is a result of the operation of the international system, rather than the deliberate choice of a given nation or of its leaders, it follows from this model that a growth of military capability, by itself, without a coupling to intent via some sort of political process, will not drive the system into war. A lesser result follows from putting Γ, γ , η (the "intent→military and overall capability" parameters) and σ equal to zero. Then military capability evolves independently of intent and overall capability evolves independently of intent or military capability. In this case, military and overall capabilities evolve linearly and non-chaotically though the intent equations are still complex enough so that the absence of the possibility of chaos in them is not obvious. If it is postulated that the absence of chaos in military capabilities is sufficient to imply the absence of crisis instability, no matter how national relationships may fluctuate, then the vanishing of the "action parameters" and the influence of the "irontriangle" is sufficient to guarantee "peace". If changes in intent do not lead to military procurements and if there are no "iron triangles" making demands on overall economic-political capabilities, peace will prevail in spite of arms races. This again seems to be a reasonable result - "reasonable " in that it concurs with many conventional (non-mathematical) analyses.

6. References and Additional Readings

Meadows, D., D. Meadows, J. Randers, and W. Behrens. 1972. *The Limits to Growth*, New York: Universe Books,

Lee, S. C., D. A. Zinnes, and R. G. Muncaster: 1991, "The Dynamics of International System Structure", University of Illinois preprint.

Saperstein, A. M.: 1994, "Mathematical Modeling of the Effects of 'Capability' and 'Intent' on the Stability of a Competitive International System", *Synthese* **100**, 359-378.

von Clausiwitz, Karl: 1832, *On War* (English Translation), 1968, A. Rapoport (ed.), New York: Penguin .

CHAPTER 7
CONCLUSIONS. PAST AND FUTURE RESEARCH

It is often the case that important international security decisions have to be made by decision-makers who have implicit, as well as explicit, criteria for making decisions. The explicit criteria may be based upon political, economic, ethical, ethnic, historical, scientific, technological, and other such considerations. The implicit factors may fall under the same headings but are usually held as generalized biases rather than rational understandings of the specific situation. Frequently the implicit factors dominate the decision process. Outside critics can rationally argue the issue, relying upon the explicit premises, perhaps prevailing intellectually . But to no avail. The role of the implicit criteria may not even be apparent to the policy makers who rely upon them! Furthermore, all of the relevant science needed to rationally decide among the dominant issues may not be well established. To the scientist at least, the undecideable questions can be openly displayed, each with its known and suspected pros and cons. However, scientist and non-scientist alike will then make the choice among these, scientifically still undecidable, factors using the same criteria used to decide the non-scientific questions. The explicit criteria, as well as the perhaps dominant, unexamined implicit criteria (biases) of the policy maker will govern the incomplete scientific choices as well.

This role of the implicit bias may make rational political discourse impossible. Without policy making based upon fully open dialogue, the resultant policies will be weak and perhaps dangerous. It would be useful to have a decision making process in which all premises, and their manipulation, are open for all to see, examine, criticize, and alter. The postulates would be general rather than *ad hoc* . They would not be contrived, explicitly or implictly, to decide the specific problem at issue. The chaos paradigm of this book is offered as such a decision making process.

As an example of the process applied to a major recurrent national defense problem of the nuclear age, consider the SDI (Strategic Defense Initiative) of President Reagan, briefly mentioned in Chapter 5. The basic policy choice is between erecting a "shield" or deploying a "sword" to defend a nation (the U.S. and/or the U.S.S.R.) against the thrust of an opponent's long-distance, nuclear-armed missiles.

The defensive "shield" would be a national missile defense (NMD): a system in which defensive missiles would be employed to destroy threatening strategic offensive missiles before they get dangerously near to their civilian or military targets. The threatened nation would presumedly peacefully relax under an impenetrable shield - over its homeland and, perhaps, over its troops and "interests" abroad. All would be secure, independently of the activities of their opponents. Such ideas have been proposed, and generously funded, repeatedly in the U.S. - during the Johnson, Nixon, and Reagan presidencies - and are currently again being strongly advanced by the Republican Congress. The defensive "sword" is nuclear deterrence - the plausible threat that any nuclear attacker, no matter how successful the attack, would be completely devastated by an inevitable nuclear riposte. No matter how severe and surprising the initial attack, a number of offensive missiles belonging to the target nation would be guaranteed to survive. These would be sufficient to wipe out the initial attacker as a functioning nation. With this option, the "peace" would be secured by mutual fear - MAD (mutually assured destruction = mutual assured deterrence). Deterrence is a strongly coupled arrangement between two or more antagonistic parties - the deterrers and the deterred; the deterrer must create a plausible threat, and the deterred must be convinced that the threatened devastating counterblow is indeed plausible. The relationship between the two depends upon the mass psychologies of both; it is the opposite of SDI's presumed self-reliance. Though often pursuing the shield option, the U.S., *de facto,* over the years has maintained and exercised the deterrence option.

The shield (SDI) option presumes that the aggressive sword(s), having been made impotent, would gradually wither away, i.e., the number of deployed strategically offensive missiles would shrink to zero, "naturally", as the defensive system grew and thrived. The sword option (MAD) assumes that the lingering fear of mutual devastation from the ever-present dangerous offensive missiles, due to accident or policy miscalculation, would lead the mutually threatened parties to gradually diminish these swords hovering over all by formal verifiable treaties. Such an ongoing treaty process has already given us several limits on offensive missiles and the ABM (Anti-Ballistic Missile) Treaty - limiting missile defense, thus holding each parties throat open to its opponent's sword, hence strengthening the MAD regime. (Needless to say, those favoring the NMD option are usually very hostile to the ABM Treaty.)

The explicit pros and cons - given a state of mutual hostility and offensive capability - are primarily budgetary and technological. Both sides are eager to get themselves out from under the opponent's offensive sword threat. Which of the options (if either) will work and how much will they cost? The offensive missile technology is well established and understood. Its costs are easily estimated. "MAD" was the operable strategy for much of the "cold war", and no major "hot war" between the principals occurred. Hence the deterrence-sword option certainly seems a viable though "unpleasant" way of maintaining the peace between the major nuclear powers. (But there have been numerous severe scares, and the next serious accident or miscalculation could be the last!) The shield's defensive missile technology - hitting a missile with a missile - is much more challenging. It requires major advances, not only in missile propulsion and guidance, but in long range radar detection, communication, and information processing. (Even further from understanding and eventual deployment are the defensive "ray technologies". These would destroy the many attacking missiles with particle or laser beams, fired from fixed ground bases or space-based satellites.) Testing is difficult and expensive (e.g., firing an offensive missile from California to be intercepted by a defensive missile fired from a remote Pacific atoll). There has been no significantly successful "one-on-one"test, and no serious suggestions have been made as to how to carry out a more realistic test - "many-on-many" in a battle-like atmosphere of disorganization, fright, and misinformation. System cost estimates have been made (and are huge!) but they can't be taken seriously until there are more serious technological answers. Lacking good explicit answers to the obvious explicit technological questions, the decisions go to the hidden premises.

One major premise underlying the shield option may be that it's easier to get one's way in the world, "to push others around", if one's homeland and important assets are secure from major threat or reprisal. Those opposing the shield option may be driven by an implicit fear that the shielded party may very well become an international bully. Another implicit difference between those favoring the shield and those accepting the sword may be that the former insist upon self-reliance in defense whereas the latter acknowledge and accept the dependence of their national security upon the desires and actions of others. One may accept/welcome the notion of international community; another may find the idea of such a community repugnant and/or deny its existence. One party may worship the possibilities of new technology whereas the other may be entranced by the possibilities of treaty negotiation - gadgets versus law, the old western "six-shooter" versus the U.S.Marshall's badge. Some may be suspicious of the social sciences, more accepting of the physical sciences; others may have the opposite preferences.

Deterrence depends upon the psychology of "plausible threat"; the "shield" is founded upon "physical tools". Finally, there are those who wish to "relax" behind the shelter of a shield. Indifferent to the problems of a difficult "foreign" world, they may also be indifferent as to the reality of the impenetrability of that shield. Others, wishing to engage the outside world, may find the prospects of a shield between them and that world to be diversionary or even dangerous, even if it were really to "work".

Seeking a model to foster decision-making in this situation, I found it easiest to avoid the technological problems by just assuming that all of the suggested technology would work. The defensive systems would do their presumed jobs, as designed, as would the offensive systems. If the model-suggested decision came out in favor of SDI, the final decision would rest upon technological tests. There would thus be added incentive to push the tests to fruition. If the decision was against SDI in these circumstances, it could be final; there would be no need for technological confirmations. The policy option question was turned into the question: what are the model stability implications of a working SDI/NMD? Would a model which manifested the goals of the NMD proponents - the gradual, natural, decay of the offensive parts of the model - be stable? If "no", adoption of the shield option would, *prima fascia*, lead to a dangerous crisis-unstable world. If "yes", further exploration of the shield option would be warranted.

The model (Saperstein and Mayer-Kress, 1988, referred to in Chapter 5) assumed that each side, the U.S. and the U.S.S.R., had an initial stock of offensive nuclear ICBMs (intercontinental ballistic missiles) pointed at the other. They also had initial stocks of defensive satellite based anti-ICBM missiles (designed to intercept and destroy launched ICBMs), and anti-satellite missiles (designed to protect their investment in offensive weaponry by destroying defensive weaponry). Each side procured further missiles of each kind, constrained by economic considerations, according to their perceptions of the threats at each time to the different elements of their defense system. (The procurement model was very similar to those developed in Chapters 5 and 6.) The many unknown model parameters were then adjusted to see if the model allowed a time evolution in which the number of defensive weapons stabilized while the offensive ICBM numbers went to the desired zero. It did. But the resultant model was extremely unstable. Starting with values for the initial stocks slightly different from each other, vastly different final numbers were obtained. These fluctuations - "extreme sensitivity to small perturbations" - imply chaos in the model world and hence, by the basic heuristic paradigm of this book, crisis instability in the real world. Those different sets of model parameters which implied a stable evolution of the system did not imply a "fading away" of the offensive threats - the goal of the SDI proponents. These model results suggest that a unilateral or a multilateral adoption of NMD, without explicit *apriori* agreements, either will not reach the goal desired by NMD proponents or is dangerously unstable.

Consider another policy-option-choice example, newly confronting the world community and completely unresolved as yet. As the world begins the long-sought-for demobilization of nuclear weapons, the question arises: what is to be done with the surplus plutonium "pits" from the demilitarized nuclear explosives? They represent enormous unrealized national investments of time, money, material and human resources. They also represent the possibility of the creation of new nuclear weapons - either openly or surreptitiously. The two answers commonly advanced seem to be "burn" then or "bury" them. Both options are currently under serious consideration in the U.S.

Combine the plutonium with other nuclear fuels and "burn" them in nuclear power reactors, thus retrieving the initial investment, supplying needed civilian energy, and destroying their ability to ever-again be used as components of fission weapons. Irreversibly turn swords into ploughshares. Or, perhaps, put the world on the dangerous course of a vastly increased trade in radioactive materials with a high potential for diversion into future weapons manufacture. Alternatively, chemically combine the fissile plutonium with inert materials, such as glass, and "bury" them in inaccessible places. But what is inaccessible for one party may be open to others with stronger desires and/or deeper pockets. What can be chemically combined can be uncombined. Even if man does not re-access the buried radioactive plutonium for re-use, nature may eventually return it to the human environment, to damage future generations. And the transformed plutonium still has to be moved to the burial sites, thus representing a still dangerous commerce. Which to choose? How to reach a decision?

The explicit criteria pertinent to this question are easily conveyed to those politically and/or technologically knowledgable. They include the energy requirements to be met by the expanded fuel source and its costs, the domestic and international environmental and security impacts of the resultant trade and use of plutonium, which may be intended or unintended, and the security and "permanence" of the final disposition. The important implicit criteria might include the fear of invisible radioactivity versus complacent acceptance of visible biological and chemical pollutants from non-nuclear energy resources. There may be a dread of the new and unknown environmental insult and a familiarity, hence contempt, for the usual ordinary environmental degradations. There might be a tilt toward a worship of the new, or a respect for old technology, a preference for physics, or a trust of chemistry. The bias might be for "common sense" (that which was learned at a very early age), or for science -perhaps never learned but culturally respectable. Or the choice might come down to NIMBY ("not in my back yard") versus WYSIMNFWRIG ("whatever you spend in my neighborhood, for whatever reason, is good"); some wish to avoid any change in their neighborhood whereas others welcome any input of new resources, no matter what their source.

Any or all of these often unseen and/or unadmitted reasons may dominate the decision, currently being considered, in spite of the openly expressed reasons advanced in the usual, informal, decision making process. It would be useful to have an alternative, *complementary,* formal process, in which all assumptions and processes are explicit. Instead of ad hoc arguments, deductions would be formally obtained from axioms which are more general than the specific case under consideration. For example, instead of postulating that reliance upon radioactive materials is good or bad, one might resort to axioms about the stability of a competitive inter-nation system. The hope is that the usual, implicit, "personal-psychological"criteria can be removed from dominating the decision process.

Thus, a model of a system of competing states might be postulated in which both weapons and energy resources are important grounds and tools of conflict. Mutual competitiveness would be enhanced by fear of the opponent's weapons stocks and his weapons potential as well as competition for decreasingly available world energy supplies. (The previous chapter has illustrated how capability and intent may influence each other.) Competition would be decreased, on the other hand, by increasing availability of energy supplies and decreased availability of weapons.

The plutonium at issue is both a potential weapons threat and an energy source, and so both possibilities would be included in the model: As one nation takes nuclear weapons out of service, its threat to the other nations of the system decreases so that they may diminish their rates of arming. This would be manifested, for example, in the reaction terms of the Richardson model (the "a" and "b" terms of Eq.1 of Chapter 3 which represent the weapons aquisition rate response of one side to the weapons stocks of the other). This "relaxation" would occur whether the weapons decline was due to either burning or burying.

On the other hand, in the Richardson Model , the constant term (e or f of Eq.1 of Chapter 3) represents the general attitude of the nation in question with respect to the rest of the world. Positive values, which lead to an increasing acquisition of arms, stem from a suspicion and distrust of the other nations of the system. A nation, dependent upon others for the energy resources necessary for its very survival, is likely to be very suspicious - towards the suppliers and other who might be competing for the same supplies. Such suspicion makes one always on guard, always arming. As its reliance upon external energy diminishes with increasing domestic energy supplies, this suspicion parameter should decrease. If these increased energy stocks come from the conversion of existing nuclear weapons ("burning the pits"), the decrease in the suspicion parameter is enhanced with an increase in the number of weapons taken out of service. Thus the choice between burning and burying manifests itself differently in the different terms of the Richardson model. This implies that the different policy choices will lead to different ranges of Richardsonian stability. Adopting, *a priori*, the explicit criterion of maximizing the realm of stability, and assuming that a Richardson-type model can be developed which contains all *important aspects* of the system (which may require a more complex model than has been so far employed), a rational, above-board policy decision on the burn versus bury question might be obtained. The same procedure may be applied to a more realistic, non-linear, model where the goal is lack of chaos rather than Richardsonian stability. Work of this kind is currently being pursued by the author.

The previous chapters have illustrated the application of the paradigm to more theoretical questions. These two examples illustrate the application of the chaos=crisis instability paradigm to pressing practical problems. One, the stability analysis of the SDI problem has been previously published and may have had some influence in the debate over the issue during the Reagan presidency. (Given the current resurgence of interest in NMD, further work along this line would be useful. The stability factors of an impending mutli-polar nuclear missile world may differ greatly - for the better, or for the worse - from those of the bi-polar world previously investigated.) The plutonium pit application is still under active investigation.

Many other previous, present, and future problems of international security, both practical and theoretical, may also be amenable to the techniques advanced in this book and their extensions. The difficulties of the method do not lie in the finding of the stability regimes of a given model - this is a purely mathematical and/or computational exercise. They lie rather in the expressing of <u>all</u> of the <u>relevant</u> variables in an appropriately interacting form while discarding those variables which are unimportant for the problem in question. The choice of the relevant variables and parameters from the many possible in any given situation requires judgement based upon experience.

Modeling of strategic and tactical security systems - the selecting and discarding of pertinent variables and parameters - to help create policy is not a new endeavor. It has been done in one form or another by military and political elites for a long time, e.g., simulations and war games - on boards, computers, or in the field. The introduction of mathematics into these fields is not novel. What is new, and developed in this book, is the realization that the specific predictive results of such models need not be the primary goals of the modeling process. An understanding of the stability of the process - whether or not a chaotic transition is possible, and, if so, where - should be the primary goal of the investigation.

Hopefully some of the readers of this book may be sufficiently intrigued by this paradigm to carry further this process. In addition to further development of the theory, applications could include the continuation of the joint Russian-U.S. missile build-down, the incorporation of the "minor" nuclear powers in this build-down, the attempts to control nuclear and missile proliferation, the incorporation of non-nuclear methods of mass destruction, the role of "minor" military powers, the role of non-state (="terrorist"?) threats, and the role of economics and the search for prosperity in the development of international hostility and threats to security. The world is not, and will not be, lacking in important threats to its security and prosperity. These should be addressed by analysts and policymakers (="statesmen"?) using the best of both new and old methods. The methods themselves require continued investigation, extension, and updating. The author hopes that the readers of this volume will apply their energies and thoughts to these important, intellectually stimulating, and emotionally rewarding matters.

INDEX

—G—

—H—

—I—

—L—

—M—

—N—

—O—

—P—